Simplicius
Corollaries on Place and Time

# Simplicius

## Corollaries on Place and Time

Translated by J. O. Urmson

Annotated by Lucas Siorvanes

Cornell University Press

Ithaca, New York

First published 1992 Cornell University Press.

**Library of Congress Cataloging-in-Publication Data**

Simplicius, of Cilicia.
[Corollarium de loco. English]
Corollaries on place and time / Simplicius ; translated by J. O. Urmson ; annotated by Lucas Siorvanes.
p. cm. — (Ancient commentators on Aristotle)
Translation of: Corollarium de loco and Corollarium de tempore, two excursus from Commentarium in Aristotelem.
Includes bibliographical references and indexes.
ISBN 0–8014–2713–4 (alk paper)
1. Place (Philosophy)—Early works to 1800. Time—Early works to 1800. 3. Aristotle. Physics. I. Urmson, J. O. II. Siorvanes, Lucas. III. Simplicius, of Cilicia. Corollarium de tempore. English. 1992. IV. Title. V. Series.
B703.S53C3713 1992
114—dc20 91–21147



Printed in Great Britain

# Contents

# Introduction

## *Richard Sorabji*

Simplicius has left us two unique and extraordinarily interesting histories, one of Greek theories of place or space, the other of Greek theories of time. They are digressions which interrupt his commentary on Aristotle's *Physics*.

### Place

Aristotle thought of place as a two-dimensional surface. My place, for example, is the two-dimensional surface of the air that surrounds me, or to avoid certain difficulties it may be better to say the inner surface of the surrounding room walls.[1] The Latin Middle Ages experienced the greatest difficulty in escaping from this two-dimensional conception. But the robust Greeks with few exceptions[2] rejected it from the start. Aristotle's pupil and immediate successor Theophrastus raised the most devastating questions about it, questions which are recorded only here by Simplicius,[3] and Simplicius himself starts by attacking Aristotle's idea.[4] Many Greeks adopted a conception of place much more familiar to us as three-dimensional space. But the Neoplatonist theories on

[1] Yet other interpretations of Aristotle are discussed in R. Sorabji, *Matter, Space and Motion* (*MSM*), London and Ithaca N.Y. 1988, ch. 11.

[2] A peripatetic precursor of some of the medieval defensive permutations is recorded by Philoponus *in Phys.* 588,6-17: we should add the idea of the *relation* of the walls' surface to myself. Cf. Thomas Aquinas *in Phys. 4*, lectio 6, described in *MSM*, 190, and Edward Grant, 'The medieval doctrine of place: some fundamental doctrines and solutions', in A. Maierù and A. Paravicini Bagliani, eds, *Studi sul XIV secolo in memoria di Anneliese Maier*, Rome 1981.

[3] 604,5-11. These questions are analysed in *MSM*, ch. 11, along with a translation.

[4] 601,25-611,10.

which Simplicius concentrates had peculiarities of their own. The Athenian Neoplatonist Proclus, for example, (*c*. AD 411-485) regarded space as a very fine body which exactly coincided with the physical world, giving us two bodies, as we might say, in the same place.[5]

One question that had already been raised by Theophrastus, so we learn,[6] was whether there was such a thing as Aristotle's natural place. Natural place was held by Aristotle to have a power (*dunamis*)[7] of its own which could account for the natural fall of rocks towards a lower place and rise of flames to an upper one. But, Theophrastus asked, might not the natural movements of the four elements rather be explained by thinking of the world as like an organism that readjusts the position of its limbs when they are out of place? The power of distributing earth downwards and fire upwards would then lie in the organism itself, not in the places which its limbs occupy. Some such conception was to appeal to the Stoics and to the Neoplatonist Philoponus.

Although Simplicius and the other Neoplatonists whom he reports favour the comparison with an organism,[8] they are not prepared to treat place as inert in this way. Four of them assign to place the power of sorting the four elements into their proper regions: Syrianus,[9] Proclus[10] and, as we shall see, Damascius and Simplicius himself. And other dynamic powers are attributed to place as well. This starts with Iamblichus' 'intellective theory', which sought to rescue Aristotle's categories, including the category of place or 'where', as acceptable to Platonists. Iamblichus' answer to Plotinus' attack was that the categories were fully applicable to the intellective world of the Platonic Forms. Indeed, the categories had their primary application to that intellective

[5] 611,10-618,7, described in *MSM*, ch. 7.

[6] 693,15-22. This and all the remaining theories are analysed in *MSM*, ch. 12, along with translations of short key passages. An earlier translation of some key passages was supplied by S. Sambursky, *The Concept of Place in Late Neoplatonism*, Jerusalem 1982, and there is a French translation of portions chosen to throw light on Proclus in A.-J. Festugière, *Proclus, Commentaire sur la République*, vol. 3, Paris 1970, 328-48.

[7] Aristotle *Physics* 4.1, 208b11.

[8] 625,15-16; 626,8-27; 627,29; 628,1-2; 629,13-18; 645,12-13.

[9] 618,25-619,2.

[10] 613,28.

world. And this can be seen, if we think of a thing's place as something which embraces it in such a way as to prevent it from dissipating and falling further way from the unity found in the supreme Neoplatonic God, the One.[11]

Not only do Simplicius and his teacher Damascius agree that place has this power of drawing bodies together.[12] But in the passages already cited, which treat the world as like an organism, they assign to place, not to the organism, the power of arranging its members. The world's members would be the four elements, the heavens and their parts. As an individual organism, I too have a place of my own, and for me too it is place that arranges my head, feet and other members. The strangeness of this conception of place becomes apparent when Simplicius infers that my unique place must move with me, when I move,[13] contrary to the almost universal conception of place as immobile.

Damascius distinguishes the positioning (*thesis*) of an organism's parts from its place (*topos*), which he regards as a measure (*metron*) of that positioning,[14] as Simplicius agrees.[15] Taking up an idea which is first assigned to Proclus,[16] Damascius compares this measure to a sort of mould (*tupos*), into which the organism should fit.[17] Simplicius, who agrees,[18] explains that the mould must be a flexible one, so that it can allow for a variety of positionings, as it does in the case of the moving heavens.[19] But Simplicius does not agree with Damascius' further idea that the measure gives things their size as well as their arrangement.[20] There are three elements in measuring: the thing measured, the physical instrument of measuring and the unit in the world of Platonic Forms by which a thing is measured.[21] The last is unextended, and this may be why Damascius regards place as

[11] 640,4-11; Simplicius *in Cat.* 361,15-20; 361,30-362,4; 362,23-6.

[12] 625,28; 631,38; 636,8-13; 637,8; 638,2.

[13] 629,8-12; 637,25-30.

[14] 625,28; 627,2; 627,14-15; 644,14; 644,23-4; 645,4.

[15] 626,3; 627,14; 630,12; 631,38; 634,2-24; 635,14-27; 638,6-17; 644,14-645,17; 774,6.

[16] 613,9.

[17] 645,7-10.

[18] 643,12-13; 773,25.

[19] 633,11-18.

[20] 644,24-5; 645,6-17.

[21] 634,11-31; 636,7-8; 636,27-637,18.

having no dimensions,[22] a third view to set alongside Aristotle's two-dimensional conception and the normal three-dimensional one.

My own unique (*idios*) place,[23] which moves along with me, and is naturally united with me and essential to me (*sumphutos*, *sumpephukôs*, *sunousiômenos*, *kat' ousian*, *ousiôdês*), [24] cannot be the only kind of place, and Simplicius sees the need to draw several distinctions. There is also such a thing as place which I share (*koinos*) with others. A shared place which I occupy, though not unique to me, may be unique to something else. It may, for example, be the unique place of the air.[25] But I can also be assigned a merely adventitious (*epeisaktos*) place.[26] Damascius had given as examples of adventitious positions being in the house, or in the market place.[27] As for the immediate place within the air which I occupy, that has no permanent existence, but perishes when I leave it. But the place of the whole air retains the ability again to fit me into some part of itself, and when it does so it will measure my extension again.[28]

Because of his peculiar conception, Simplicius thinks of place not as an extension (*diastasis*) and hence as something in the category of quantity, but as a substance or essence (*ousia*) which is extended.[29] In fact Simplicius has the problem that a thing's place is hard to distinguish (*dusdiakritos*) from its form (*eidos*) or essence (*ousia*).[30] But in fact form gets its own orderliness from place.[31] Although Simplicius claims that Iamblichus, Damascius and he are in the tradition of Theophrastus, because Theophrastus suggests that things may have a place dictated to them by nature,[32] his dynamic view of place in fact reverses Theophrastus' position. For Theophrastus had suggested that

[22] 601,16-19.
[23] 629,8-12; 637,25-30.
[24] 625,14-17; 627,16-17; 629,8-12; 633,11-12; 638,26.
[25] 629,8-12; 637,25-30.
[26] 627,16-17.
[27] 625,14-18.
[28] 632,9-26.
[29] 623,19-20.
[30] 629,13-19; 630,30; 638,26-7.
[31] 630,24-30.
[32] 639,10; 642,17.

the arrangement of the universe might be due not to the power of place, but to the natural form (*emmorphos phusis*) of the cosmos.[33]

## Time

Simplicius sees Proclus and Damascius as original in their theories of place,[34] and he thinks there has been progress in the treatment of time too, because until his own day, no one had solved the paradoxes recorded by Aristotle about the existence of time.[35] These paradoxes are not introduced until near the end of the Corollary on Time,[36] but the Neoplatonist theories were partly shaped by them, and can conveniently be introduced by reference to them. The paradoxes had been transmitted to the Neoplatonist Iamblichus in the version of pseudo-Archytas.[37] The latter wrote probably late in the first century BC, but the Neoplatonists were taken in by the pretence that this was the work of the original Pythagorean Archytas, who lived around 400 BC, before Aristotle. The paradoxes were therefore taken to have been derived by Aristotle from the Pythagoreans, instead of the other way around.

In Aristotle's version,[38] the first paradox declares that time does not exist because its parts do not exist: the future does not yet exist and the past exists no longer. As for the present, it is not a part of time, because it is a sizeless instant with no length. If you think the present has a length, you will find that part of that length belongs to the past and part to the future, and the present cannot overlap with these.

The second paradox complains that we cannot remain stuck at the same present instant as the people of ten thousand years ago, so the present instant must cease to exist, but when? It cannot cease to exist while it is in existence. It cannot cease to exist at the very next instant, because, as Aristotle rightly sees, sizeless instants are never next to each

33 639,18-19.
34 611,12; 625,2.
35 795,33-5.
36 795,27ff.
37 786,4-10; Simplicius *in Cat.* 352,29-30; *in Cat.* 353,13-15.
38 Aristotle *Physics* 4.10.

other, any more than are geometrical points on a line. Finally, it cannot remain in existence for, say, a second, until some later instant, for then it would have had to stay in existence at all the infinitely many intervening instants.

Aristotle rejects both paradoxes, but does not tell at any point in his discussion how to solve them. I believe it is clear from other works that he would solve the second paradox by saying that the present instant never is ceasing (present tense) to exist, but that we can use the perfect tense and say that it *has* ceased. It will have ceased at any instant we like to pick after its occurrence, however close.[39] Aristotle's insistence on using the perfect tense 'has occurred' rather than the present is highlighted by Simplicius in a related context at 797,8. As for the first paradox, Simplicius conjectures that Aristotle would solve it by saying that a past and a future do not have actual existence.[40]

The first *Neoplatonist* solution of this paradox, however, is provided by Iamblichus. Like Plotinus before him[41] and Proclus,[42] Damascius[43] and Simplicius[44] after him, he distinguishes between a higher and lower time.[45] We have already encountered Iamblichus and Damascius referring to a higher kind of place. The higher time is not a Platonic Form, because Plato described time as being a mere image of the eternity of the Forms.[46] But it is raised above the world of change, and this gives Iamblichus one part of his solution. Admittedly, the higher present is indivisible, or simple, as Urmson translates it, and so is not a part of time. But it is immune from paradox, because the higher time is not subject to division and flow; so it cannot be said of it that some is no longer and some is yet to come. As for the lower time, which is subject to flow, a Neoplatonist will not mind admitting to its unreality, although not for the reason alleged in the paradox.

[39] See Aristotle *Metaph.* 3.5, 1002a28-b11, with discussion in R. Sorabji, *Time, Creation and the Continuum* (*TCC*), London and Ithaca N.Y. 1983, ch. 1, pp.10-12.

[40] 800,21-4.

[41] 790,30-792,19.

[42] 795,4-11; Proclus *in Tim.* 3.32,7-16.

[43] 787,30; 788,9-10; 788,29-32.

[44] Simplicius *in Cat.* 355,8-14.

[45] 787,10-11; 793,3-7; 793,12-22; 793,28-9; 794,21; Simplicius *in Cat.* 352,14; 354,14; 355,8-14; 355,27-356,1.

[46] Plato *Timaeus* 37D.

For the present in flowing time is not a mere indivisible instant (this is the meaning of 'now' at 793,22-3), as the paradox alleges, but is a whole stretch of time between two instants.[47]

Damascius offers a more elaborate answer which draws on a certain solution to Zeno of Elea's paradox of the half-distances. According to that paradox, you cannot reach any destination, because in order to get there, you would first have to go half way, then half the remaining distance, then half the remaining distance again, *ad infinitum*. A solution, probably Stoic, is recorded by Sextus Empiricus.[48] Why should you not reach your destination by disappearing from one spot and reappearing at your destination, without delay, and without ever having been part way in between? May not motion always occur in this cinematographic way, without our being able to detect it perceptually? Since the Stoics are not atomists, these leaps will not be of some standard atomic length, but are divisible – they may in other words be across as short a spatial distance as you like. On the other hand, there is a sense in which they are also indivisible, because, if we think of time instead of space, we see that they take no time at all.

What Damascius wants to do is infer that, if the celestial clock moves by leaps like this, then time too will progress by leaps which are both divisible and indivisible,[49] although he does not succeed in showing how the combination of divisibility and indivisibility which might be true of motion could equally be true of time. A further source of confusion is that he gives the name 'leap' not, as we might expect, to the instantaneous transition from one time to another, but to the intervening period between two transitions.[50] Damascius thinks he can solve the Aristotelian paradoxes by insisting that the present is not an indivisible instant (or now,

[47] 787,10-11; 793,22-3; Simplicius *in Cat.* 354,9-33; 355,8-14; 355,29-356,1. The solution is analysed in *TCC*, ch. 3, pp. 37-41, with a translation of key passages. A German translation of the entire Corollary is supplied by E. Sonderegger, *Simplikios zur Zeit*, *Hypomnemata* vol. 70, Göttingen 1982.

[48] Sextus *M* 10.123-42. The solutions of Simplicius and Damascius, along with this background in Sextus, are analysed, with translations of key passages, in *TCC*, ch. 5.

[49] Damascius, *Commentary on Plato's Parmenides* (*in Parm.*) = *Dubitationes et Solutiones*, ed. Ruelle, vol. 2, 241,29-242,15.

[50] Damascius *in Parm.*, = *Dub. et Sol.* 2.242,24-6.

796,27-9), as the paradox-mongers try to show,[51] but is one of his leap-like periods of rest. Provided that he can insist on the indivisibility of these periods, he will be able to resist the paradox-mongers' attempt to split the present period up without remainder into past and future portions. It will rather be a genuine *part* of time, and therefore one part of time at least has existence.

But this is not all he does in order to solve the paradox of the parts of time. He combines the idea of the leap-like present with another important idea, that we should consider what we mean by talking of existence in connection with time. For time to exist is for one part to be coming into being after another, not for all the parts to exist at once.[52] Consequently, we can go further than saying that at least part of time exists. The whole of time exists, and exists simultaneously in reality,[53] thanks to one part coming into being after another. The parts in question are the leaps. Damascius keeps repeating that time does not progress by instants, but by leaps.[54] And these leaps are parts of time, and can be said to have being while they last.[55] Because the leap is, in some sense, indivisible, we do not have to suppose that it in turn depends for its existence on the coming into being of its parts.

Damascius also offers to solve Aristotle's paradox of when the present instant ceases. His answer is that there is no time at which it ceases, because to suppose that there is would be to imply that time stands in need of a second time. But Aristotle denounced the idea (*apephêne*, or better, used to deny, reading *apephê* at 800,1-2) that there was a process of process, and it would lead to a regress to suppose that a measure stands in need of a measure.[56] This may leave us wondering whether we can avoid the other horn of Aristotle's dilemma, that the present instant of ten thousand years ago will not have ceased. But Damascius also makes a further point applicable to both paradoxes, that the paradox-mongers

51 796,27-9; 799,18-30.
52 797,36-798,4; 799,8-18.
53 775,33-4; 798,4.
54 796,34-797,5; *in Parm.* = *Dub. et Sol.* 2.236,15-18; 2.242,5-6.
55 799,19-20.
56 799,35-800,16.

are wrong to treat the present instant as if it enjoyed actuality.[57]

Returning to the paradox of the parts of time, Simplicius wants to keep some of Damascius' solution, while rejecting other aspects. He wants to conclude that time exists, but not that the whole of it exists simultaneously in reality,[58] and he does not like the leaps. In fact he turns something that Damascius says against him. For Damascius claims that it is somewhat of a falsification to divide a flowing thing into segments like the past, present and future, or days, months and years.[59] In that case, Simplicius complains, it must also be a falsification to segment it into leaps. What he would like to retain from Damascius is rather the idea that the existence of time consists in one thing's coming into being after another.[60] But Simplicius is in difficulties, for what are the things which will come into being? Not segments of time, if time cannot be segmented, and Simplicius would be as reluctant as Damascius to allow time to exist through the coming into being of sizeless instants.

Simplicius finishes by appealing to another idea in Damascius, who says that instants are not actual but only potential divisions of time,[61] and equates that with the divisions being made only in thought.[62] Simplicius agrees that we can segment time only in thought.[63] But if the instants which segment time exist only in thought, then, as Damascius himself indicated, the paradoxes which turn on time actually being segmented cannot arise. This is how Simplicius thinks Aristotle meant to deal with them.[64]

The Neoplatonist theories of time are the main ones described in Simplicius' Corollary, but not the only ones. Considerable space is given to a philosopher eight hundred years earlier than Simplicius: Strato, the third head of Aristotle's school. This is because he raised detailed objections

[57] 799,30-5.
[58] 775,33-4; 777,11.
[59] 798,5-799,8.
[60] 797,27-36.
[61] 798,9-10; 799,34.
[62] 798,10-11, an idea ascribed at 748,23-4 to an earlier philosopher, Alexander of Aphrodisias.
[63] 797,32.
[64] 800,21-4.

against Aristotle's definition, and set up his own rival definition of time as the quantity (*poson*) in actions.[65]

## Acknowledgements

The present translations have been made possible by generous and imaginative funding from the following sources: the National Endowment for the Humanities, Division of Research Programs, an independent federal agency of the USA; the Leverhulme Trust; the British Academy; the Jowett Copyright Trustees; the Royal Society (UK); Centro Internazionale A. Beltrame di Storia dello Spazio e del Tempo (Padua); Mario Mignucci; Liverpool University. I further wish to thank David Furley and Lucas Siorvanes for their comments on the translation and Paul Opperman and Lucas Siorvanes for compiling the indexes. I am grateful to Ian Crystal and Paul Opperman for their help in preparing the volume for press.

[65] 788,36-790,26.

# Textual Emendations

### Corollary on Place

| | |
|---|---|
| 605,32-33 | Supplying quotation marks. |
| 606,33 | Reading *autoi* instead of *auto* with Spengel. |
| 610,7 | Reading *skopein* instead of *skopei* with MSS aF. |
| 611,16 | Deleting quotation marks before *anangkê*. |
| 611,21 | Deleting quotation marks after *metaxu*. |
| 614,29 | Reading *oute* instead of *ouse*. |
| 617,11 | Reading *monon* instead of *monên*. |
| 617,16 | Supplying e.g. *eirêmenê* as Diels suggests in apparatus. |
| 617,39 | Replacing question mark by full stop after *legetôsan*. |
| 620,4 | Reading *holôs* instead of *holos*. |
| 620,5 | Reading *somata* instead of *asomata* (cf. 619.34). |
| 629,4 | Reading *proslogizontai* instead of *prologizontai* as Diels in apparatus. |
| 630,10-11 | Replacing comma by question mark after *einai* and question mark by full stop after *topôi*. |
| 631,5 | Reading *topos* instead of *logos*. |
| 638,1 | Reading *auton* instead of *auto*. |
| 638,18 | Deleting quotation marks. |
| 638,18 | Deleting comma before and after *phêsin*. |
| 638,23 | Deleting quotation marks. |
| 640,34 | Reading *merôn* instead of *monôn* as Diels in apparatus. |

### Corollary on Time

| | |
|---|---|
| 777,29 | Reading *houtô deixoi* instead of *egô deixô* as Diels in apparatus. |

| | |
|---|---|
| 779,36 | Reading *hupotithetai* instead of *apotithetai* with MS a. |
| 781,24 | Reading *lambanein* instead of *lambanôn* as Diels in apparatus. |
| 784,29 | Supplying e.g. *kai tou mê hama holon ekhontos to einai*. |
| 786,12-13 | Reading *kinasios tinos arithmos ê kai katholô* instead of *kinasios tis arithmos ê kai katholô*. Cf. *kinêseôs tinos* at 786,13 and 787,30. See also *Simplikios: Über die Zeit*, Anhang IV, by Erwin Sonderegger in *Hypomnemata*, Heft 70, Göttingen 1982. |
| 789,35 | Reading *polun* instead of *pollên*. |
| 790,23 | Closing quote after *ousi*. |
| 791,30 | Reading *auto* instead of *autên*. |
| 793,11 | Closing quote after *praxeôn*. |
| 795,16 | Reading *philopthonian* instead of *philophonian*. Possibly *philoponian* as Sambursky and Sonderegger; cf. 795,34. |
| 797,35-6 | Text seems corrupt but the sense is clear. |
| 798,22 | Reading *enestôs* instead of *huphestôs*. |
| 799,20 | Reading *deiknusi* instead of *deiknuousin*; cf. *deiknusi* at 799,24. |
| 799,29-30 | Omitting *einai* (29) and *kai to peratoumenon* (30). |
| 800,1-2 | Denounced (*apephêne*), or better, reading *apephê*, used to deny. (Ed.) |

# Simplicius

# Corollaries on Place and Time

# Translation

# Translator's Note

The translation follows the text of *Simplicii in Aristotelis Physicorum Libros Quattuor Priores Commentaria*, edited by H. Diels in *Commentaria in Aristotelem Graeca*, vol. 9, Berlin 1882, except as indicated in the footnotes and Textual Emendations. The marginal numbers give the page and line of that text. The reader should be aware that there was no equivalent of quotation marks in the original Greek. They were added by the editor and are not always certainly correct.

While some knowledge of Neoplatonic thought is an advantage, the translation is intended to be as comprehensible as possible to anyone who understands English. Uniformity of translation of the key terms has, however, required some sacrifice of natural idiom. In rendering technical terms the translation mainly follows the edition of Proclus' *Elements of Theology* by E.R. Dodds. The main departure from tradition, and from Dodds, is that *kinêsis* is translated 'process' instead of 'motion'. Those who find such expressions as 'motion in respect of quality' congenial can easily so emend the translation.

The footnotes are by Lucas Siorvanes unless otherwise indicated in parentheses at the end of the note.

# Corollary on Place

## Contents

| | |
|---|---|
| 634,11 | Reply to fifth objection. |
| 637,11 | Reply to sixth and seventh objections. |
| 637,33 | Reply to eighth objection. |
| 638,15 | Reply to ninth objection. |
| 638,18 | Reply to final objection. |
| 639,10 | Anticipations of Damascius: Theophrastus and Iamblichus. |
| 640,18 | Explanation of the disagreements discussed. |
| 642,15 | Best theory is that of Damascius. |
| 644,5 | Some truth in all theories considered. |
| 644,20 | Failure of Damascius to distinguish between extension as position and as magnitude. |
| 645,18 | Discussion of place ends. Will examine Aristotle on the void next. |

# Simplicius: Corollary on Place

601,1 Such is the character of Aristotle's account of place. It contains many difficulties and offered many lines of examination to those who came after him. So I wish to set out the objections brought against him and to bring to light the cause of his faulty argument about place. But even then my account does not seem to be complete. For there have been other opinions about place since Aristotle, an examination of which he would have handed down to us if they had arisen before him. So he would approve of them also being examined. Also the theory stating that place is an interval,[1] briefly tested by him but approved by the most distinguished of his successors, seems to me to be worthy of more extended consideration. If I were able myself to contribute to the articulation of our thoughts about place I think that Aristotle would countenance my daring, since he has provided the basis himself. So if I shall seem to exceed the office of a commentator, let those who notice it blame the difficulty and complexity of the problem.

Let it be known, then, that of those who have written about place[2] some have held the theory that place is body, others that it is incorporeal. Proclus, the Lycian philosopher, for example, held that it is body. Of those who said that it is incorporeal, some said that it is altogether unextended, others that it is extended. Of those who said that it is altogether

[1] The term *diastêma* is translated principally as 'interval'. For the Neoplatonists it usually refers to interval expanded in three dimensions, viz. space-extension. The Neoplatonic *diastêma* has its context in the metaphysical theory of 'procession' (*proodos*) from the One to the Many. Being extends from unity and wholeness to plurality and particularity. This extension is like the radiation of light from its source – hence the popular name 'emanation'. It is also like the multiplication of numbers from unity. *diastêma* is the filling-out of fragmented being. Cf. 613,5; 773,21.

[2] Simplicius presents the theories on place in the manner of a Platonic division. A diagram may be helpful:

unextended some said that it is a substrate for bodies, like Plato, who calls the basic stuff place; others said that it brought bodies to completion, like our own Damascius. Of those who said that it is extended, some said that it is two-dimensional, like Aristotle[3] and all the Peripatetics, some that it is three-dimensional. Of these, some said that it is absolutely homogeneous and sometimes remains without any body, like the followers of Democritus and Epicurus,[4] some that it is an interval, always containing body and fitted to each, like the well-known Platonists and Strato of Lampsacus.

But since Aristotle's views are fresh in our memory, let us examine them first. One should first pay attention to two points that Aristotle ran together, that the heaven does not change its place and that neither it nor the universe is in a place. Furthermore, he does not say that the fixed area alone is motionless and not in place, as his expositors say,[5] thinking to diminish placelessness, but also the whole heaven, since he

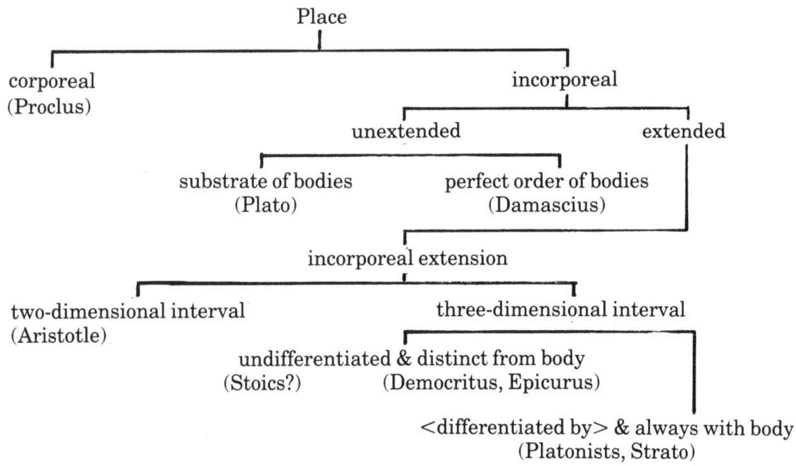


[3] If place is the limit of the container, or the surface immediately adjoining what is in place, place is a surface and so two-dimensional. (J.O.U.)

[4] 'Undifferentiated' (*adiaphoron*) is the description of place not divided up by the presence of a sensible body. On the dividing effect of sensible bodies on immaterial entities in Neoplatonism, see Proclus *Elements of Theology* (Dodds), proposition 80. Homogeneous, universally extended body may refer to the Stoics, though Simplicius does not mention them (but cf. Syrianus *in Metaph.* 84,27-85). That place, or space, is distinguishable from the bodies that occupy it, and can notionally or actually exist without them, was generally characteristic of the Atomist theories of Democritus, Epicurus and their followers.

[5] Aristotle's expositors (*exêgêtai*), viz. the Commentators.

saw it as one. That he thinks that only things moving rectilinearly both change their place and are in a place could be confirmed from many places. To save space, it will be 602,1 sufficient to quote one or two passages: 'It is not the heaven that is place, but the extreme portion of it that touches the moving body. For that reason the earth is in water, which is in the air, and the air is in the aether, and the aether is in the heaven, but the heaven is not in anything further' [*Physics* 212b18].[6] We must beg to differ from Alexander, who says that fire has been omitted here and that the planetary sphere is called aether. Also before this passage Aristotle says: 'I call the contained body that which moves locally' [212a6]. But he calls contained that which is in place. If, then, the contained body moves locally, i.e. is in place, then either (1) change of position is something different from circular motion and signifies rectilinear motion, as even the expositors agree; in which case those things would be in place which are subject to rectilinear motion, which are properly called moveable, since they are not in everlasting motion like bodies in circular motion, but have the capacity to move; or else (2), if one were to use passages from the last book of this work and *de Caelo* to equate local motion with change of place in such a way as to include things in circular motion as locally moved,[7] one would clearly have to admit that the fixed stars and the whole heaven were in place; but this Aristotle clearly forbids by admitting nothing outside the container.

We have already cited the passages in which Aristotle distinguishes local motion from circular motion. The passage in which he treats it as the common genus of change of place is this: 'There are three kinds of process – change of size, change of state and change of place, which we call local motion' (*Physics* 260a26). He later divides local motion, saying: 'Everything in local motion moves either in a straight line, or a circle, or in a combination of these' [261b28]. From this it is crystal clear that also according to him circular motion is change of place. But what changes its place is also in a place.

[6] Aristotle *Phys*. 4.5, 212b18-23: ... *kinêtou sômatos peras êremoun*. *peras êremoun* omitted by Simplicius. *houtos d' en tôi aitheri* Aristotle; *ho de aêr en tôi aitheri* Simplicius.

[7] See also Proclus *Elements of Physics* (Ritzenfeld) 2, propositions 4-6, 17, 21.

So it follows that the whole heaven and the fixed stars are in a place. How, then, in his lectures on place, can he say that the fixed stars and the whole heaven and this whole universe, things defined as being fixed, are not in place? Also he added the common explanation of these facts, when he said: 'So that body is in a place which has something outside it containing it; that which has not is not' [212a31]. This is a reasonable saying if place is the limit of the container. Alexander relies more on this and does not accept the previously cited passages, saying that circular motion is not change of place because it is not displacement as a whole. He explicitly writes as follows: 'For if it does move, still this is not change of place; for it does not exchange its place since it is not contained.' But it is plain that what is in circular motion is in local motion and that local motion is a motion. But if it is a change of place, as
603,1 the forementioned passage said, it is plain that it is also in place. For that which changes its place is also in a place, and is admitted by them to be so, seeing that it was principally from change of place that place was discovered. But if they say that it does not change its place, let them say what species of motion or alteration circular motion is; it is certainly not substantial, nor qualitative nor quantitative change. Aristotle says most clearly that circular motion is change of place in *de Caelo*, where he says: 'Every change which is of place, which we call local motion, is either in a straight line, or circular, or a combination of these' [*de Caelo*, 268b17]. No less clearly he assigns change of place to perpetual bodies, saying that 'the only change perpetual bodies undergo' is change of place [*Physics* 260b29]. He adds the explanation elsewhere, saying: 'For the changing thing changes its nature least' when the change is local motion [261a20]. 'For in this way only it changes none of its nature, as in alteration it changes qualitatively and when increasing or diminishing it changes quantitatively.' From these quotations one would say that it is clear that neither truly nor consistently does he seem to say that only bodies moving rectilinearly change their place and are in a place; but in accepting the consequences of this definition of place, whatever they may be, he runs its risks. How can he say that 'there would have been no inquiry into place if there had been no change of place' [211a12] and add

'and our main reason for thinking that the heavens are in place is that they are always in motion', and yet so clearly deny that the heavens are in place, saying: 'But the heaven, as has been said, is not anywhere as a whole, nor in any place, at least if no body contains it' [212b8]?

How can he say that that is in a place which moves with local motion? For that which is in place does not invariably change its place. The earth itself, as he says himself, is in water as a place and remains stationary. He himself made this clear when explaining the difference in the rest of what is above and of what is below. 'For the latter is ever at rest,' he says, meaning it remains in the same place motionless, 'but the extremity of that which is in circular motion is at rest as being unchanging' [212a23], having regard to its rest in essence in this case. Further, if things in something moving are in it as in a vessel rather than in a place, as he himself explains saying: 'So when something moves and changes place within something moving, like a boat in a river, it uses its container as a vessel rather than as a place' [212a15], then the planetary area will be within the fixed sphere as in a vessel and similarly each sphere of the planetary area in that containing it.

Moreover, all the sublunary, which changes by local motion and which he says is properly in place, this too is within the limit of the heavens as in a vessel, since it is within the hollow surface of the sphere of the moon; for the sphere of the moon is also in motion. Where, then, is place, i.e. what things are properly in place? For both air and water are in motion, while individual bodies in general are in flux in the air or water. It should be known that Theophrastus also in his *Physics*[8] raised difficulties about the account of place given by Aristotle, such as the following: that body will be in a surface; that place will be in motion; that not all body will be in place (for the fixed heavens will not be); that if the spheres be taken together the whole of the heavens will not be in place; that things in place, themselves remaining unchanged, will no longer be in place if the things containing them be removed.

[8] = Theophrastus fr. 146 FHS&G, *Theophrastus of Eresos*, eds W.W. Fortenbaugh, P.M. Huby, R.W. Sharples and D. Gutas, Leiden 1991. I am grateful to R.W. Sharples of Project Theophrastus for the full reference.

But let us see if the definition of place given is consistent with the postulates about it previously accepted. We all say that a place contains the things in it. It is clear that this is not put forward as a unique property of place, but in virtue of the fact that to be in a place is to be in something or other. For we say that everything that is in something is contained by that in which it is. In what sense of 'containment', then, do we say that place contains? Is it insofar as it pervades the whole of what it contains, or is it insofar as it envelops it from the outside? But if, on the one hand, it pervades the whole, how could it be the limit of the container? Yet, if it envelops from the outside, in that case a cloak would also be place, something that even those who use metaphors would not care to say. Again, that which contains as the limit of what is outside contains either the whole or the surface. But it is impossible for it to contain the whole. For the container must be either greater than or as great as what is contained. Now the limit of the container is as great as the surface of what is contained; but the surface cannot be the whole of what is contained. Further, the surface is not in a place, as Aristotle himself agrees, if what is in place is that which is as such separate and in motion, as he says, and if the surface is the limit of a body in place. But perhaps somebody will believe that a whole is in a place in virtue of its surface. However, that, too, is impossible; for if something does or undergoes something in virtue of a part, it is necessary that the part should do or undergo it pre-eminently. For example, if somebody kicks in virtue of his foot, it is necessary that his foot should do that pre-eminently, and if somebody is whitened in virtue of his surface, it is necessary that his surface should undergo that pre-eminently. But the second of the presuppositions was that the place was as great as what is in the place. But how is it possible for a surface to be as great as a body? If it were, a point would be as great as a line, and a line as a surface, and thus a point as great as a body. What could be more absurd than that? For it is clear from the fact that he says that what moves by local motion is in place that body is naturally in place.[9] But it is not the surface but the

[9] Reading e.g. *ek toutou hoti legei* instead of *ek toutou legein* at 605,1. (J.O.U.)

body that moves *per se*. However, containment seems to be understood as some sort of envelopment and not as the pervasion of the whole, and to acquire the notion of equality via that former sense.[10]

Further, he says, we take above and below to be distinguishing features of place [cf. *Physics* 205b32]. So just as there is something that is above all else, beyond which there is neither place nor anything in place, so there must be something that is below all else. If, then, what is above all is that which surrounds, whether as the convex surface of the fixed heavens, or that of the aether, if only the sublunary is in place, then what is below all else will be the middle of the universe, and this is the centre point and its immediate surroundings. So, it seems, the place of the earth, which is below all else, is not the limit of water or air, as Aristotle made clear, saying: 'On this account the earth is in water, water is in the air and the air in the aether' [*Physics* 212b20].

But perhaps someone will say that 'above' and 'below' each have two senses, one as used of place, the other as used of what is in place, as Aristotle made clear when he said: 'what contains from the middle is the lower limit, as is the middle itself, but what contains from the outermost is above, as is the outermost part itself' [212a26]. For he himself agrees both in this work and in the *de Caelo* that the middle is what is below and what contains is above; and, since the universe is spherical, it is clear that, as what is below is the centre, so the above is the limits of straight lines thrown out from the centre. In that case, what would the centre be? For it is not a place, if place is a surface and container and equal in size to what is in place; but nor is it in a place, if what is in place has a surface and is moveable by local motion. But the place above is neither the concave surface of the lunar sphere nor the concave surface of the fixed heavens, but, if anything, the convex surface. For it is at this surface that the extremities of the lines from the centre terminate. Further, if the lunar sphere moves in a circle, and if the sphere of fire also moves in a circle with it, as Aristotle agrees in the *Meteorologica*

[10] According to Simplicius the dimensional equality of the container to the contained seems to be understood by Aristotle as that of the envelope to the enveloped. See *Phys*. 4.1, 209a5-15, 209b1-12.

[341b18],[11] then this above will, according to his own definition, be a distinguishing feature not of place but of a
30 vessel, as he defined it. 'For when', he says, 'something moves and changes place within something moving, like a boat in a river, it uses its container as a vessel rather than as a place. For a place', he says, 'is naturally motionless' [*Physics* 212a16]. Therefore, in his definition also, he says that 'place is the immediate unmoved boundary of the container' [212a20].[12] So if what he calls the above is not a distinguishing mark of place, but of a vessel, clearly the same applies to the below.

606,1 Aristotle stated as a final postulate that bodies of all sorts naturally moved to and remained in their proper places, and that this fact constituted the above and below. So that if earth moves downwards, and the centre is below, earth will not move towards the limit of water or of air, but towards the
5 centre. Therefore that is the place of the earth. It does, indeed, move thus, for heavy things move directly downwards from all directions. This would not happen, the earth being spherical, unless everything had an impulse towards the centre. But the centre, which has no parts, is not the limit of the containing body. Further, he himself says that the place of the whole and the part is identical. In the *De Caelo* [cf. 294a10ff.], where he
10 is inquiring into the cause of the earth's immobility in the middle, he named not the limit of the air but the middle as the explanation why if it were now removed elsewhere it would move there. Also, in the third book of the *Physics*, where he criticises Anaxagoras because he says that the boundless holds itself firm within itself [205b1ff.], he himself says: 'But the heavy remains in the middle and the earth is in the

[11] Aristotle's firesphere just below the moon was exploited by the Commentators, particularly the Neoplatonists, for the criticism and rejection of the celestial aether. If the firesphere can rotate naturally like the heavens, then there is no need for aether as a separate element to constitute the heavens. The overall context of the debate was the natural motion of the elements in relation to proper place. Simplicius' more detailed account, including on Xenarchus, is found in *in Cael.* There he mentions Ptolemy, Plotinus and Proclus (37,33-4). Briefly, S.'s references to Ptolemy are *On the Elements* (not extant), *Optics* (partly extant). For Plotinus, *Enn.* 2.1.8, 2.2.1; Proclus, *in Platonis Timaeum Commentaria* (Diehl) 3.111,22-112. Later Philoponus, *de Aeternitati Mundi contra Aristotelem*, ap. Simplicium *in Cael.* 34,5-37, employed the same argument to conclude that the celestial realm is perishable, like the sublunary, in accordance with the Christian doctrine of the end of the world. See n. 19 below.

[12] Supplying quotation marks at 605,32-3. (J.O.U.)

middle' [205b15]. If, then, the moving object both moves towards its proper place and remains in it, it is clear that the middle must be the place of the earth. Further, if every body naturally moves to and rests in its proper place, in order to gain its natural state, and each thing is preserved by its natural state, how is it that water moves towards the limit of the air? For, as Aristotle agrees, the less quantity of water is mastered and transformed by the greater quantity of air.[13] With what aim, then, does water move towards what destroys it? But it wishes to be close to the earth and to be supported by it. Therefore, if you take away the underlying earth, it will move again, until it finds more earth, and it deserts the air containing it and its limit as not being its proper place. Further, if the limit of the air were the place of water or earth, the water in the air and clod of earth, being surrounded on all sides by air and possessing their proper place, would not move. But if it moves because, if it gains unity it will be less liable to suffer harm, it aims at its own unity and not unity of place. Further, if places are proper to bodies, why are the places occupied by a single body of different kinds? For if somebody standing in a river rises above the water with the upper parts of his body the place of one and the same body will be the limit both of the earth on which he stands, and of the water, and of the air, the water containing the lower parts of the body and the air the upper parts.

Now Theophrastus and Eudemus[14] themselves accept the immobility of place also as a postulate,[15] but Aristotle makes an addition to the postulate, saying: 'so that place is the immediate unmoved boundary of the container' [*Physics* 212a20]. Yet he says that what is in place all changes its place. But now everything is contrariwise, if place is the limit of the container; for earth and water are at rest, but the limit of the air that contains them is in motion. Further, if place has to be motionless, either Aristotle says falsely that place is the limit of the heavens that touches bodies in motion, or, if that is true, the heavens ought to be motionless in order that their

[13] The greater part of air masters the lesser part of water, Aristotle *GC* 331a25-35; 334b-335a.

[14] = Theophrastus fr. 147 FHS&G. I am grateful to R.W. Sharples of Project Theophrastus for the full reference.

[15] Reading *autoi* instead of *auto* at 606,33 with Spengel. (J.O.U.)

limit also might be motionless. For if they are in motion their limit will not be motionless. But it is clear that he holds that the heavens are in circular motion, and the perceptible facts bear witness to it. So it is necessary either to loosen up the postulate that says that place is motionless, or else not to say simply that place is the limit of the container. But, further, our intuition does not only require place to be immobile but also in its nature more enduring. For that very reason Aristotle himself characterises the above and below by their constant rest and says: 'the below remains constant and the extremity of the circle of the heavens remains unchanging' [212a23]. If, then, place ought to be more enduring in its nature how could the the limit of the air be the place of the earth? For air is more easily changeable than earth.

But I think that it is right to attend carefully to this also, that the defender of this view of place will say not only that the heavens are neither in place nor change their place but the same also of complete wholes below the moon. For it is clear that according to him they will not be essentially in place, if that is essentially in place which is in local motion, i.e. changing one place for another. But that they will not be in place even contingently is clear from the following: if they are in place because their parts are in motion, and the whole is comprised in its parts, consider the consequence; just those parts will be in motion locally and will be in place essentially which are cut off and separated from the complete wholes; but a whole does not consist of these.

It is possible to introduce many other considerations. But what will contribute most to the orderly exposition of these doctrines about place is, I think, to discover the principal bases that lead to a conception of place that encounters so many objections. First, then, I think that the conception of place as a container is to blame and secondly the assumption about that which essentially changes its place. For while it is the general conception of place that it contains that which is in the place, the term 'containment' is taken in many senses; for everything which is in something seems to be contained by that in which it is. Aristotle accepted the idea of a container as that which covers up, according to which, as I said already, a cloak would be a place, it not being a natural whole that is

contained by it but a whole in virtue of its surface, since the whole is defined in that way. He reveals that he had such a conception when he says that the heavens are perhaps the universe [*Physics* 212a23] because it is the furthest out of all 608,1
things and for that reason contains all things. Thus he took it that the contained is a whole in virtue of the surface bounding the mass and so said that the limit of the container was the limit of body as a whole. However, a place seems to be a certain space and a receptacle of a whole, not a covering of it.[16] I think that much was contributed to such a conception by the fact that the parts seem to be in a place not *per se* but as belonging to a whole, if, indeed the assumption about containment was not rather the cause of this conception. For if the parts had, themselves, a place, then place would have spread through them, cohering with that whole place as they with their own unity. But that the parts were not *per se* in place was taken as evident because continuous parts do not exist *per se*. But, first, perhaps the parts have an intermediate nature between the totally inseparable elements and the totally separate individual things,[17] so that one could not say to which class they belonged or could see them simultaneously in both ways, as existing *per se*, or not. Further to that, if, though they are continuous, we said that the parts were in different places, it would be truly absurd; but, if they are in a continuous place, how much more would it follow that, as with what is in place, so with place, whole corresponding to whole, part to part, and continuous to continuous, separate to separate? For in that way place will really be like a receptacle, and one should rather have thus distinguished a vessel from a place, on the ground that a vessel merely encloses but place permeates the whole, rather than on the ground of the one being moveable, the other immoveable.

[16] Place as space (*khôra*), literally. room or area; and as a receptacle (*hupodokhê*) of the whole. See Plato, *Timaeus* 50B-51B, 52A. It is Aristotle who identified the Platonic space-receptacle with his own 'matter' (*Phys.* 4.2, 209b).

[17] The 'totally differentiated monads' (*diakekrimenôn monadôn*) would be distinct individuals as wholes; the 'totally undifferentiated elements' (*adiakritôn stoikheiôn*) the intrinsic constituents of definition. The parts may be regarded ambiguously, according to Simplicius' exposition; on the relation of parts to elements, Damascius *Principiis* (= *Prime Principles*) (Westerink), 2.174-76 (= 1.196-7 Ruelle); Proclus: whole from parts, whole in parts, *El. Th.* propositions 66-9, and A.C. Lloyd, *The Anatomy of Neoplatonism*, Oxford 1990, 82-4.

Concerning the assumption about that which essentially changes its place, I say that, as was his regular custom, he wished to make his approach to place starting from observable things and, restricting himself to these, was constrained to accept this assumption. For in fact it was change of place that gave us the conception of place, and, of things that change their place, those things especially which move from place to place as wholes. For seeing, as Aristotle himself said [*Physics* 208b1], that where previously there was air there was now water, we came to the conception of the common receptacle, which we called place. So since it was from these especially that we recognised place, it was these that he accepted as being in place.

But someone who has not yet fully grasped Aristotle's conception might say that, even if it was from these things that we gained a conception of place, still it was also from circular motion, as Aristotle himself says in the last book of this work, that we came to have the concept of the unchanging cause of all things. But surely we do not, on that account, say that the participation of that cause in the circular motion is changelessly, primarily and properly the infinitely powerful cause of all
609,1 things?[18] What necessity, then, is there to dwell on those things alone that brought the notion to mind? But since we completely so behaved and took those things to be what properly changes its place, so that we even counted what moves in a straight line as all that is in place, it became necessary finally to mark off that which was *per se* in place as being such as is bounded by something else and to say such things as Aristotle says. So, when what contains a thing is not discrete but continuous and adjacent, things are not said to be in it as in a place but to be a part within a whole; but when a thing is discrete and adjacent it is in the immediate boundary of its container. For what changes place for place will inevitably take over a place that it did not occupy before. Therefore, he says that things that change through locomotion and growth and diminution are of that sort, but things that take over a place either take it over when it is empty of body (though he lays it down that every place has a body just as every body has a place)

[18] Infinite in power (*apeirodunamos*), Proclus *El. Th.* propositions 84-6.

or, alternatively, will take over the place of what adjoins it and is bounded by it. So there must always be something adjacent to such a body and containing it.

This, I think, is the explanation why he assumed that what is in place was inevitably contained by something else from outside, but did not reverse the argument. For he does not say that what is contained by another thing is in place or changes its place, unless it also changes one place for another. Thus the conception of place arose from these considerations. It is for this reason also that he does not say that the planets are in place, although they are contained by each other and by the fixed heavens. But why talk of the planets? For I do not think that Aristotle will say that the whole totalities of elements are in place. For if he says that everything in place also changes its place, then if the moveable body is in place and locally moveable, i.e. in a straight line, since circular motion is not change of place, it is clear that he holds that only things travelling in straight lines change their place *per se* and are in place, because only these things exchange their place. But such in the sublunary sphere are the parts that are fragmented from their proper totality. These totalities, however, are either at rest or are in circular motion,[19] unless, indeed one may conceive of them as growing or diminishing; for then they seem to move locally by expanding or relinquishing their place. So if nothing celestial is in place, and of the sublunary the parts are not because they are continuous and the totalities are not because they do not change their place, nothing natural can be in place, except possibly plants and complex organisms.

I think that the definition having this character was assembled from these starting points. He seems to collect the same from his fourfold division – for he says that place is either matter or form or the limit of the container or the interval between the boundaries; but it is none of the other three so it must be the limit of the container. It is likely that

[19] The elements are either at rest or are moving circularly: this is a late antique repudiation of Aristotle's rectilinear motion of the four elements: Plotinus *Enn.* 2.2.1; Proclus *in Tim.* 2.11,27-12,8. For studies, L. Siorvanes, 'Proclus on the Elements and the celestial bodies: physical thought in late Neoplatonism', PhD diss. London 1987, II A.5; C. Wildberg, 'John Philoponus' criticism of Aristotle's theory of ether', Berlin and New York 1988, II 5.4.

this too followed from his notion of containment, through which there will seem to be nothing in between for those who posit containment as mere envelopment. It also followed perhaps from the definition of body, which says that body is three-dimensional and that what is three-dimensional is body.[20] It will be more appropriate to consider[21] whether this is exact or not a little later in the discussion of dimension. At present it should suffice to know that anyone who posits that everything three-dimensional is body and that no body permeates another body must repudiate the notion of place as the interval between the boundaries of a container.

But we shall learn that this four-fold division is not invariably maintained when we examine other theories of place beyond these four. What I would say at present is that there are many other types of the relationship of being within such that they are neither that of matter nor form nor of interval between, but which are none the less not that of being a limit of a container. For neither the whole in which are the parts, nor the parts in which the whole consists, nor the genus, nor the time is of that type, nor is anything else except for things functioning as vessels and in general suited to envelop. Aristotle himself does not seem to take it as a division, but rather as an enumeration of the views of place that come to mind. But, if it is not taken as a division, the investigation seems to aim merely at acceptability and not at demonstration. That the arguments brought against the contained interval are not probative we shall quickly see in our discussion of interval.

But now, having said so much concerning that theory of place that we have discussed and others that we shall discuss, I shall turn to other issues concerning how, by fixing our attention on the special requirements of place and judging the discussion of it with reference to various theories, including Aristotle's, we have all inevitably fallen into many absurdities. For why on earth does the whole need a place if the parts

[20] Definition of body as the three-dimensional: S. will examine this shortly, 611,11. Proclus will exploit Aristotle's definition of place as three-dimensional arguing that since only things of the same kind can be related, then the containing place must be of the same kind as the contained body. Cf. Sextus Empiricus *Adversus Mathematicos* 2.19-28; 3.19-22, 77-91; Proclus *in Tim.* 2.77; 2.113,12-32; 3.329,30-2.

[21] Reading *skopein* instead of *skopei* at 610,7 with MSS aF. (J.O.U.)

are not in a place? If bodies need a place *qua* bodies, surely it is not the case that the whole is a body but the parts incorporeal. If, as wholes, bodies need a place, why then are not the heavens in a place and also the whole universe? Yet Aristotle seems to say that those things alone need a place that change their position as wholes and take over a place they were not in before, just as he says that perpetual things have no need of time, but only things existing in a part of time, because the whole of time does not exist, except part by part. So if time is necessary for the exchanges of position of things that cannot remain in the same place nor move without exchange of position, he would plausibly say that those things alone were in place. Notice how the discussion has focussed, not on the whole nature of the subject of enquiry, but on what is close at hand for us and familiar.[22]

What, however, does it mean to say that even such things as can remain stationary or move in the same position, like the heavens, are in the same position? Is it as being inside themselves? But Aristotle says that no body is in itself [*Physics* 210a25ff.] since a thing and what is in it are different. And even if the heavens be said not to move since they do not exchange one place for another, still they remain in the same position and are said to have constant position not from their essential nature but as occupying the same place. Thus the cause of the objections is the failure to set out in advance the role of place completely. It will be demonstrated later how the divine[23] Aristotle well discovered that to which he gave attention, even if he did not encompass all questions

[22] Discussion focuses first on familiar, immediate and concrete things, and then moves to the general and abstract. This is the path of the investigation into Nature outlined by Aristotle in the proemium of the *Physics* (1.1, 184a17-27).

[23] The personal appellation 'divine' has an interesting place in the Neoplatonists. Usually it is reserved for Plato (Proclus *in Tim.* 3.9,22; Simplicius *in Cael.* 377,27), and for great Platonist metaphysicians, such as Iamblichus (Simplicius *in Cael.* 564,11), but also for Pythagoras, as the antecedent of Plato in the Golden Chain of higher truth (Iamblichus *Vita Pyth.* (Klein) 29.162,3-4; Syrianus *in Metaph.* 81,31). Aristotle is called *daimonios* (Syrianus *in Metaph.* 115,23-5; Proclus *in Tim.* 1.6,22; Simplicius *in Phys.* 1359,5). In Greek *double entendre* it means 'ingenious', and a 'messenger mediating to deity'. According to the relative position of Aristotle and Plato in the Neoplatonic curriculum, the metaphysics of the former reaches only to *nous*, but of the latter to the Good and One. So A. is as an angel to P. as god; or as the clever to the wise. Apart from an oblique reference to Aristotle and Plato as 'divine men' (*in Cael.* 87,26-7), Simplicius seems to be the only Neoplatonist to elevate Aristotle in this one instance.

about place, through being the first to investigate it. After Aristotle's theory, let us set out, if we may, that theory which was propounded with great originality by Proclus, the Lycian philosopher, who was the teacher of our teachers. Of all of whom we know, he was the only one who elected to say that place is a body.[24] Proclus, then, accepted Aristotle's postulates about place and also took over his fourfold division for the investigation of place. He said[25] that it was necessary that it should be matter, or form, or the limit of the containing body, or an interval equal to what is in the place between the limits of the container. For if place is neither something in the body nor something about it, it would not be able to exchange its place, since nothing either in it or about it was undergoing a change. He said that what was in the body was form and matter and about it were the limits of the container and the interval between them.[26] So having shown that place was neither matter nor form, using the same arguments as Aristotle, and having demolished the view that it was the limit of the container on account of the absurdities adduced against that view, he concluded that place is the interval, thus providing a link to the proof of his own opinion.

Since Proclus expounded his opinion clearly and expertly, it would perhaps be better to listen to his own words, which are as follows: 'It remains, accordingly, that if place is not the form of what is in a place, nor its matter, nor the limit of the container, then that interval between the limits of its

[24] This fragment is our chief source for Proclus' innovative theory of place and space. The other equally important, but usually neglected, source is Proclus *in Platonis Rem Publicam Commentarii* (Kroll) 2.198-202, in which Pr. gives a more detailed analysis of the properties of place as a body, including that it is indivisible (*adiaireton*), continuous (*sunekhes*), most-immaterial and most-divine (*aülotaton, theotaton*). Indeed I think there is a high probability that Simplicius wrote this part using Pr.'s *Commentary on the Republic* as source: the latter contains fully and explicitly many of the refs that are obscure in S., e.g. the Chaldaean Oracles, place and light, the World Soul, apparitions in aether, and the light-dichotomy (615,28-616,3). Further, Pr. seems to refer to a separate monograph dedicated to place, *peri tou topou* (*in Remp.* 2.199,22; and cf. scholia, 380,90). It is likely, however, that the original source for both fragments was Proclus' *Investigations of the Objections of Aristotle to Plato's Timaeus* (*Episkepsis <en tôi> tôn pros ton Timaion Aristotelous antirrêseôn*), which is attested, *inter alia*, at *in Tim.* 2.278,27; 279,3. For two studies that utilise the *in Remp.* fragment see L. Schrenk, 'Proclus on space as light', *Ancient Philosophy* 9, 1989, 87-94; 'Proclus on corporeal space', *Archiv für Geschichte der Philosophie* (forthcoming).

[25] Deleting quotation marks before *anangkê* at 611,16. (J.O.U.)

[26] Deleting quotation marks after *metaxu* at 611,21. (J.O.U.)

container is the immediate place of each thing. The total cosmic interval of the whole universe is something different from this interval and clearly the interval is something or nothing. If it is nothing, local motion will be from nothing to nothing, though every change involves something that is, and the natural places will be nothing. But, if it is something, it must be either corporeal or incorporeal. But it is absurd for it to be incorporeal, for place must be equal to what is in the place. But how could a body and anything incorporeal be equal? For equality is found in quantities, moreover in homogeneous quantities, as lines with lines, surfaces with surfaces and bodies with bodies. Therefore a place is a body if it is an interval. If it is a body, it is either mobile or immobile; but, if mobile, it must necessarily change its place, so that place will over again require a place. But that is impossible, as Theophrastus and, indeed, Aristotle agree.[27] For at any rate he says that a vessel is a moving place and that place is an immobile vessel, since place is of its nature immobile. But, if place is immobile, it is either indivisible by the bodies that enter it, so that a body permeates a body, or it is divisible, as air and water are by things that come into them. But, if it is divisible, when the whole is split the parts will move to either side of that which splits it. So, in the first place, body will also be mobile, if its parts move; but it has been shown to be immobile. Next, when the parts are split, we shall investigate where that which split them went. For yet another interval will be found between the parts of that which was split which received the splitting agent, into which it moved and is said to be in a place, and so endlessly. Place, then, is an indivisible body. But if it is indivisible it is either immaterial or material. But if it is material it is not indivisible; for all material bodies, when other material bodies move into them, suffer division by them, as when our bodies fall into water.[28] Only immaterial bodies are such as to be divisible by nothing. This is necessarily so; for every immaterial body is impassive, but no divisible body is impassive.[29] For division is something a body

[27] = Theophrastus fr. 148 FHS&G. I am grateful to R.W. Sharples of Project Theophrastus for the full reference.

[28] The argument against material interpenetration can be found in Syrianus *in Metaph.* 612,16-24 (*CAG* 6.1).

[29] Every immaterial body is impassive but no divisible body is impassive, Proclus,

suffers which destroys its unity, since you would find no happening other than division which takes away the continuity of the continuous *qua* continuous. So, to draw together all that has been demonstrated, place is a body that is immobile, indivisible and immaterial. If so, it is very obvious that it is also more lacking in matter than all other bodies, both those in motion and even those among them that are immaterial.[30] So that, if, of these, light is the most simple (for of the other elements fire is the least corporeal, but light less so than fire), it is clear that place should be light, the purest among all bodies. Accordingly', he says, 'we should envisage two spheres, the one of light, the other consisting of many bodies, equal to each other in bulk. Situate the one about the centre and put the other inside it, and you will see the whole universe within place, moving within the immobile light, being itself immobile as a whole, in imitation of place, but having moving parts, in order to be in that way inferior to place.'[31] Having said this, he confirms his account from Plato, who says that that light, akin to the rainbow, which is
613,1 mentioned in the *Republic* [616B], is place, and also from the

---

*El. Th.* propositions 80 and 46-8.

[30] The moving immaterial bodies are the celestial bodies. Materiality and immateriality are relative properties in Proclus: 'immaterial by comparison to the changeable matter' (*hôs pros tên metablêtên hulên*), the sublunary are truly 'enmattered' (*enula*), *in Tim.* 2.46,18-20. Only the supracosmic entities, such as soul, and the intelligibles are immaterial in an absolute sense. Such immateriality, however, refers to the gross matter of bodies, not to matter as the substrate power of the universe. R. Sorabji, *MSM*, 114-16, alleges among other difficulties for Proclus' treatment of immateriality that he sometimes exploits a different non-relative sense of 'immaterial' as 'lacking in prime matter'. But I think he sticks to the relative sense. Proclus *does not reject* matter as 'universal receptacle': the Platonic analogue of Aristotle's prime matter. 'I mean ... by immaterial and enmattered in relation to the most gross matter (*aülon kai anulon pros tên pakhutatên hulên*) ... We learn that matter penetrates through all the world, as the Gods say; for this reason he [Plato *Tim.* 49A 51a] proceeds to call it receptacle of the universe' (*in Tim.* 2.10,4-9). See also *in Tim.* 2.46,19-27; 3.122,18-20; and L. Siorvanes, 'Proclus on the Elements and the celestial bodies: physical thought in late Neoplatonism', PhD diss. London 1987, 212, 278-9. On S.'s view of Pr.'s material heavens and the common substrate, cf. 615,15-19 below.

[31] Hierarchy of being and motion in Proclus: intellect = unmoved; soul = self-moved; bodies = moved by another. Cf. *El. Th.* propositions 14-24. Further, for the Neoplatonists the whole is metaphysically superior to parts. Light fulfils the conditions of corporeality, immobility, changelessness, interpenetration, *in Remp.* 2.200-1.

Chaldaean Oracles, citing what is said about the soul from the primal source:[32]

> giving life from on high to light, fire, aether and the universes.

This is that light which is above the sphere of fire, a monad prior to the triad of the spheres of fire, aether and matter. This, he says, is what first received the perpetual realms of the gods and what displays the direct visions within itself to those who are worthy. He says that, within this, formless things are given their form, according to the oracle. Perhaps he would say that it is called place (*topos*) as being, as it were, a mould (*tupos*) of the whole body of the universe and as making the inseparable separate.[33]

With regard to this, he sets himself the problems how a body will permeate another body and whether place is lifeless or partakes of soul. But he says that it is impossible for it to be lifeless, both because it is superior to the things within it that have souls and because the oracles say that it was given a soul before all other things.[34] But, if it has a soul, how can it be immobile? He solves the first problem through the impassibility of immaterial bodies. 'For', he says, 'the immaterial body neither exerts nor suffers pressure. For what suffers pressure has a nature capable of being affected by things that press on it. But it neither divides nor is divided, being impassive, so that it is not possible to adduce the paradox that the whole

[32] Part of a Chaldaean Oracle repeated in 616,1 and 617,5. The complete version is found in Proclus *in Remp.* 2.201,14-16 = Chaldaean Oracle fr. 51 (des Places). Note that 'aether' lies under fire (and above the material 'worlds' of air, water and earth): for Proclus this is additional evidence that 'aether' is not a celestial element, but a form of *sublunary* air and fire, as Plato had said in *Tim.* 58D1.

[33] Shapeless things take form and extension with procession (on which see n. 1). Proclus *in Platonis Cratylum Commentaria* (Pasquali) 32,12-14: 'The Gods tell us to consider the extended shape of light (*morphên phôtos*). For although it is without shape above (*anô amorphôtos*) it becomes shaped (*memorphô menê*) through the procession (*proodon*).'

[34] <Cosmic> place/space was given a soul prior to all other things. The Chaldaean Oracle may be from Porphyry, who, according to Proclus *in Remp.* 2.196,23-30, was the first to suspect (but not to say) that space is the primary vehicle of the World Soul (*kosmikês psuchês okhêma prôton*). The supposition was that souls could not be wholly disembodied, so they were given vehicular bodies, which, I think, may be the way of individuating souls separable from matter. On the luminous vehicle, Pr. *El. Th.* propositions 207-9. On the astral body, E.R. Dodds, *Proclus, the Elements of Theology*, Oxford 1963, appendix II, 313-21.

will permeate the smallest part. For, if it is naturally indivisible, it could not be divided into parts equal to the smallest. But if that cannot be, nor could the whole permeate that part.' He solves the second problem by saying that place is given a soul by the soul from the primal source itself,[35] having a divine life; and it is stationary, being in its essence self-moving but not in actuality. 'For', he says, 'if we assign to the soul a twofold power of self-movement, one in essence, one in actuality, and if we call the one immobile, the other in motion, what is to prevent us from saying that place has such a soul and lives unchangeable in its nature, but that the universe has a life of actual self-movement? But if you should wish to contemplate the actual motion of place, you will see that it imparts motion to those moving bodies which unfold the parts of space as intervals, since they cannot be in the whole of place and place cannot be present in all of them with each of its parts. In this respect place is an intermediate in relation to soul, which causes motion without being extended. For life as such seems to impart motion, while place, which participates in life first of all things, imparts motion[36] in its own parts and alone displays change of place, causing each part of what is in motion to strive to be in the whole of it, though it cannot occupy the whole beyond the part because of the special nature of interval. For everything that strives to be something, but falls short of so being through a defect of nature, strives to be that which, through its weakness, it cannot be. For', he says, 'the unchanging but bodily life had to be intermediate between the incorporeal and unchanging life, such as that of the soul from the primal source, and the
614,1 changing, bodily life.[37] But', he says, 'it seems to me that the cardinal points of the whole universe are fixed in it as a unity.

[35] Primary source (*pêgaian*) of soul: occasional Neoplatonic technical term for the World Soul, which is characterised as a self-constituted *henad* and is therefore divine. However, insofar as the origin of created things, including soul, is the Intellect, the primary source of soul is the intellective power (corresponding mythologically to the Great Mother, Rhea: Proclus *Theologia Platonica* (Saffrey and Westerink) 5.37).

[36] *didonai kinêsin*: the technical terms are identical to those used by Philoponus for the so-called impetus theory in dynamics, cf. Philoponus *in DA* 154,26-8 (*CAG* 15); *contra Proclum* (Rabe), 268,5-10; *de Opificio Mundi* (Reichardt), 198,24.

[37] This is the law of mean terms in operation: soul (incorporeal and unchanging), space (corporeal and unchanging), body (corporeal and changing): thus space is the bond and bridge between matter and the intelligible.

For, if the oracles say that the cardinal points of the material universe are fixed in the aether above it, correspondingly we shall say, ascending, that the cardinal points of the highest universe are seated in that light. But perhaps', he says, 'that prime light is the image of the paternal abyss and therefore, like it, will transcend the universe.'[38]

These and similar things are what Proclus says about place. At this point, let us examine that demonstration of his by which he concluded that place is that kind of body. First let us call to mind what was said above as we object to that fourfold division. There I showed, as I thought, that it was not a division but an enumeration of opinions about place, since there are other types of account of place. If it is not taken as a division it clearly is not true to say that if place is neither matter nor form nor the limit of the container it remains that it is the interval that is equal to that which is in the place. I allow the claim that the concept of interval is now taken differently from the way in which Aristotle conceived it. For Aristotle assumed that the interval was unambiguously incorporeal, as he himself makes clear when he says 'but what is between them is regarded as empty' [*Physics* 212a14]. But Proclus' account is indefinite, so that he made a subdivision of intervals into the bodily and the incorporeal. But, quite apart from this, how could what follows be true, to the effect that the interval of place could not be incorporeal? He says that, if place lies equal to the body in the place, body and the incorporeal could not be equal to each other, because the equal is found in quantities that are homogeneous, and so the interval of place could not be incorporeal. But, if we say that things are equal which both measure the same, and the measures are of the intervals and not of the volumes of the bodies, why is it not true to say that a three-dimensional object can be equal to another three-dimensional object? For it is homogeneous with it in that respect.

A surface could not be homogeneous with or[39] equal to a

[38] 'Paternal depth' (*patrikos buthos*), also *aduton*: a Chaldaean expression for the cloud of unknowing that the mystic encounters before the Ineffable One (*arrhêton hen*). Neoplatonists, including Proclus, make it equivalent to the Platonic Intelligible, cf. Pr. *in Crat.* 57,25; *in Tim.* 1.312,7-8; 2.92,8, insofar as it is opaque to discursive reasoning, but accessible by direct knowing.

[39] Reading *oute* instead of *ouse* at 614,29. (J.O.U.)

body, being different in species because of one dimension being omitted. But why could not something that has length, breadth and depth be homogeneous in just that respect to a thing possessing the same dimensions?

But, if he held, as Aristotle did, that absolutely no such thing as interval existed, he ought to have proved it or taken it as already proved. Or, if he concedes that it is place, but denies that it is an object, because it cannot be equal to the body in place, I am puzzled how that is correctly stated, unless 615,1 he has envisaged this interval similarly to those who take it to be void. But in the void there are not three dimensions, nor interval as part of nature or as form, but merely an absence and privation. He himself, in fact, judges place to be an incorporeal natural thing. If he had not, it would be easy to refer him to the arguments of those who reject the void.

But how does he prove that the interval is an immaterial body? He says that if it were material it would be divisible and changeable. Why, then? Are not the heavens agreed to be material, and do not he and Plato and the oracles teach that they are material? Yet they too are impassible and indivisible. What, then, prevents place, like the heavens, from being material but an indivisible and impassive body having some superiority over other bodies, as he himself claims for it? Perhaps he counts as material only the sublunar region, which is really subject to change, and denies that place is like that; but, as I have said, nothing prevents him from saying that it is material in the way the heavens are.

If this is the case, in what way will place be superior to the immaterial firmaments, which themselves have a place? Proclus says himself in his letter to Aristocles that even these are material in a certain way – not, indeed, as sublunar things that change in their common matter, nor as the heavenly bodies that are composed of the four elements and therefore need a common substrate, but because they, too, change their place and contain potentiality through not being everywhere as a whole.[40] But, in the first place, it would be hazardous to

[40] The heavens are material, where matter = the nurse of Becoming: the sensible world cannot manifest at once all the forms as wholes, therefore it brings them into being in sequence. That the heavens are material was also Philoponus' conclusion, but for entirely different reasons, viz. agreement with the Christian doctrine of a perishable universe.

speak against the oracles, which say that this universe alone is material; secondly, there could be something other than matter signified by being in place, not *qua* being in motion, but because *qua* extended its parts differ from each other, each being differently situated. But why do I say this, when he himself has predicated immateriality, which is superior to the changeable and material, of place? As I have said, the heavenly region could be like this. Finally, he says that, if one must posit that place is the most immaterial of bodies, and fire is more incorporeal than the other elements, but light more so than fire, it is clear that place should be light. If, however, light is a species of fire, as we learn from the *Timaeus*,[41] and if the species is not superior to the genus, then, so far as this goes, light would not be superior to fire.

The above is what I would say, briefly, against the demonstration. Against the other supporting arguments, I would say that the one based on light being said in the *Republic* to be like the rainbow[42] seems altogether obscure. It could refer to the luminous vehicle of the Soul of the All, as Porphyry interpreted it, or to something else. But if the oracles say of the soul from the primal source,

> from on high giving life to light, fire, aether, universes, 616,1

they well confirmed by dichotomous division[43] that the light is something beyond the empyrean, the aetherial and the material, and that it is superior to them all. Also an orderly enumeration makes this clear.

Perhaps light is the monad of the triad of universes, as Proclus agrees. But it is a monad, not as being the place of the three, but as being a universe prior to them. For there must be a unit before three, just as there is a triad before the sevenfold. It may be that they[44] called this unit light as being

[41] Light a species of fire: paraphrase of Plato *Tim*. 58C.

[42] Light like a rainbow: Plato *Rep*. 616B.

[43] The source for S.'s unclear dichotomous division is Proclus *in Remp*. 2.201,21-9. There Pr. explains that the primary distinction between light and the other elements is that between things unmoved and moved. So the Chaldaean arrangement consists of one element which is not moved and unchanging, light, and three others which are moved, viz. fire, aether (as an intermediary) and the material 'worlds' (air, water and earth as a group) (L. Siorvanes, ibid., 198).

[44] i.e. the oracles. Possibly 'he', i.e. Proclus. The Greek could mean either 'he' or

the flower of the empyrean firmament and as being what is common to appearance, that is, to the perceptible, that is, to the corporeal totality. For every body is perceptible, even if different ones are appropriate to different senses. Perhaps when the oracles called the empyrean the first corporeal universe they gave the name 'light' to the whole incorporeal array that transcends the corporeal universes and is illuminated by the soul from the primal source.

However, it would be well said that place first received the perpetual realms of the gods, but I do not think that it follows that place is a body. For, if it were an interval, what prevents that also from being assigned to the divine domain, and the more so since its being is more akin to the incorporeal? That the imprints of signs and other divine apparitions appear in place is, most importantly, not altogether compatible with the oracles that say that they appear in the aether and not in light.[45] Further, they, too, could be imprinted in an interval; for they do not need a corporeal substrate in order to become visible, but, rather, an interval and an impression.

As for there being nothing paradoxical in a body pervading another body, an immaterial one pervading either a material or an immaterial one, one might accept it on the basis of Proclus' demonstrations, but mainly from divine doctrine, since it says that the empyrean pervades the aether, and the aetherial the material; and, since the intellective triads and hebdomads all have an originative role and, through this, have their domain in the empyrean, they proceed together with the universes, just as the empyrean pervades them all.

But I find it repugnant that the spheres should be incomplete and be mere vaults, not complete spheres, if they did not penetrate to the centre. There will not be any connection between the centre and the circumference of each sphere if it is not joined to it by something homogeneous. Also

'they'. (J.O.U.)

[45] = Chaldaean Oracle fr. 142 (des Places), in Proclus *in Remp.* 2.242,11-12. That the apparitions are in aether, not in light, is a reference to the secondary 'pneumatic' vehicle of the soul (cf. n. 34), which, unlike the primary luminous vehicle, is a semi-material compound made of elemental 'envelopes'. See Proclus *Pl. Th.* (Saffrey and Westerink) 3.19,3-9. On the distinction between 'aetherial' vehicle, meaning tenuous, and aether as an elemental medium, see L. Siorvanes, ibid. 190, 201-5. On Hecate, the goddess of Mysteries, Chaldaean Oracle fr. 221 (des Places), in Philoponus *de Opificio Mundi*, 202,13-14 (Reichardt).

the universes will not be solids but hollow vaults. So what can the oracles mean when they say

> the father formed seven firmaments of the universes?[46]

But if someone does not find this sufficient evidence that immaterial bodies pervade both each other and material 617,1
bodies, let him examine on this point also him who says that place is body.

How is it that, while the oracles said inclusively that both light and fire and aether and the material universes were given life by the soul from the primal source

> from on high giving life to light, fire, aether, universes

he himself says that light is immediately given life by the soul from the primal source, but not yet the other universes?[47] And what does he mean by saying that light is given life by it? Does he mean that it becomes the soul of light? But he would not say that, for how could the source of souls be the soul of some one thing? Or does he mean that the other things have their own individual souls, but light alone[48] receives the illumination from the soul from the primal source? However, they[49] do not say that the empyrean has a soul, but call it intellective, and say that souls begin at the aether, unless indeed a soul watches over every body and there is some analogy in the universes between the intellective, the psychic and the physical. But if he says that the cardinal points of the empyrean firmament are fixed in the light – for he does not mean the points of all the universes, as the analogy he sets out[50] shows – will not light stand in the same relation to the empyrean as that to the aetherial region and that to the material region? So, if these regions are not the places of these less perfect points, even though the points are fixed in them,

[46] = Chaldaean Oracle fr. 57 (des Places). The 'Father' is Chaldaean for a creator deity.

[47] This whole argument refers back to 614,1-6.

[48] Reading *monon* instead of *monên* at 617,11. (J.O.U.)

[49] Possibly 'they' are persons unspecified; if they are the oracles, Simplicius has made a not uncommon grammatical error. (J.O.U.)

[50] Supplying e.g. *eirêmenê* at 617,16 as Diels in apparatus. (J.O.U.)

then for that very reason light cannot be place. How could it be, since light is the image of the whole paternal abyss and not of the ground of the intelligible, as Proclus would sooner have it? For the abyss transcends the triadic division of the universes and is said to divide the Whole into three. On it the monad of the triad of universes would depend as one universe, not as the place of the universes. For, if the universe were not one before it was three, how would place be connected with the three and be first of all?

That is what one might say about such analogies. But what Proclus has to say in displaying the agreement of such an hypothesis with the postulates about place will accommodate it no better to the view that place is body than to the view that it is an incorporeal interval. Similarly, the problems that he raises and solves as special to place are common to practically all hypotheses. So I shall try to deal collectively with such problems at the end of the discussion of place.

But now it occurs to me to ask those who propose that place is an immaterial body whether it is the place of bodies because it is body or because it is immaterial. But they would not say that it is because it is body, but because it is an immaterial body. And yet, while they say that the firmaments above the material universe are immaterial, they do not say that they are place but that they need a place. So, if it is not itself in need of a place either as being a body or as being an immaterial body, but is a place for other things, let them say as being what else it is so.[51] Secondly, if it is not in need of a place because it is immaterial, it seems that it is matter that primarily has a need of a place. However, who would say that matter was in place by its definition? But if someone should say that the material object was, why is it that immaterial things are in place? Thirdly, and most importantly, one might bring the same objection against these men as against the former, that they neglected to inquire into the special character of place, whatever it may be, and into the service it does to the things in it. If that is not discovered, I think that it is impossible to learn whether place is a body or incorporeal. For consider, if *qua* body each thing needed a place, is it

[51] Replacing question mark by full stop after *legetôsan* at 617,39. (J.O.U.)

possible that place should be a body and not itself stand in need of a place?

Among those who say that space is an incorporeal interval there is much agreement and also disagreement. In so far as they all say that that which receives bodies is incorporeal, they seem to agree with each other. But in so far as some see it ontologically as nothing other than the absence of body, and also say that it has a capacity to receive bodies not as a natural quality but because, as being nothing, it also does not get in the way of things inside it,[52] while others say that it has a natural capacity, and think that it does not have this receptive character contingently, but essentially, they differ. There are also those who say that place is superior to bodies.

Again, of those that say that place is void, some say that it is boundless and exceeds bodies in boundlessness, and for that reason receives different things in different portions of itself indiscriminately, if one can speak of portions with regard to a boundless void. The ancient natural philosophers associated with Democritus seem to have held such a view about it. But others make it equal in extent to the corporeal universe, and say that for that reason it is empty in its own nature, but is for ever filled with bodies and is only notionally seen as existing in itself. Many of the Platonic philosophers were of this persuasion, and I think that Strato of Lampsacus held this view.

Among those who held that place has a form and power superior to that of bodies I would place the great Syrianus, the teacher of the Lycian Proclus. In his commentary on the tenth book of Plato's *Laws* he wrote as follows about place.[53] He calls it an interval and says: 'It is this that puts this and that kind of body in its proper place within its own segments and divisions, which it got from the different principles of things in the soul and from the illumination of the creative forms. It makes itself in some parts the proper place of fire, to which fire is said in the *Timaeus*[54] to move by its nature; in other

[52] The sense of this clause is clear, but it contains too many verbs. (J.O.U.)

[53] Syrianus on place, cf. *in Metaph.* 86. Syrianus' view of a superior place is shared by Proclus. Iamblichus' view of place as a controlling power (see Introduction) would count too.

[54] *Timaeus* 63B. (J.O.U.)

parts it makes itself the proper place of earth, towards which earth naturally moves and where it rests when there. Therefore', he says, 'now also all moving and stationary things continue each in a place. But neither the motion nor the rest in respect of interval is subject to the nature of bodies nor is it brought about by that nature.'

Against those who put forward that place is empty interval, whether bounded or boundless, the impossibility of ascribing to a void the capacity of place will suffice, as Aristotle also says, and as one can prove from the natural motion of bodies towards their own places. For what might be the capacity or special features of the void, viewed as void, so that some of it should have an affinity with some sorts of body, some with others, or so that one body would be seated at the middle, another at the outside? If it were boundless, how could it be bounded by a middle and an outside, which are boundaries? If it were bounded, it would have its limit merely contingently through body, and a middle and outside similarly. For such a void would not have a boundary or a difference between its middle and its outside, being more or less the negation of body and nothing other than what is not. That such a thing is nothing is clear from the fact that, if one notionally takes it away, another empty interval will appear where the other was, just as we leave it there if we remove a body. Continuing to do this without limit, we shall not cease to replace void by void for ever, because our fiction was to no purpose and truly void. Aristotle also will well oppose those who hold such a position.

Someone might attempt to prove that the empty interval is also place as follows, making use of the same presuppositions about place: place, that needs to encompass the whole of the body in that place, must have the same dimensions as the body, and be equal to it in all the three dimensions. What is three dimensional is either body or an incorporeal interval; but if it is body it is inevitably itself in place *qua* body. For how, *qua* body, will it differ from a body in place? But if place needs no place it cannot be a body. So place is therefore an incorporeal interval, having the capacity to contain bodies. That is why Aristotle, voicing common conceptions, said: 'So place has three dimensions, length, depth and width' [*Physics*

209a4].[55] So place cannot be a body.

But what do they mean by saying that it is not body? For we say that everything three-dimensional is body, but not every body is a natural body. Thus we say that there is a mathematical body, which we also call a solid, which is defined by its three dimensions. But we do not claim that mathematical bodies are in a place. So that, even if somebody strove to show that the three-dimensional is body, it still will not be such as to need a place, but is in between a theoretical, ideal body, which has one incorporeal reality and is endowed with a soul, and the material, natural body which is primarily called body. For, taken as a whole,[56] what is primarily called body is not so defined in virtue of its dimensions, for, if it were, mathematical bodies would not have been called bodies;[57] in addition to its dimensions, the primary physical body has a certain fullness by which it differs from mathematical bodies.

So there seem to be two magnitudes, equal to each other, that of place and that of body, the one having a receptive capacity in addition to dimension, the other a capacity to fill the receptacle, so that, whatever the dimensions of the space, to that extent it can equal the body in the place and, in so far as it has no body, to that extent it can receive a body which is not impeded by anything. This, it seems, is the image in the universe of the intelligible empty gap (*khaos*), the receptacle (*khôra*) of bodies, which is called the gap (*khaos*) from the vocable '*khô*', which signifies the verb 'contain' (*khôrô*).[58]

That such an interval was not thought to be a mere empty notion can be seen also as follows: we say that what changes its place moves from somewhere to somewhere; but from somewhere to somewhere must contain an interval between. But the interval between either belongs to the body in motion, or to another body through which it moves, or to the air, or is itself something essentially incorporeal. For there is no other

[55] At 601,20 Simplicius says that Aristotle held place to be two-dimensional, presumably because it is the inner surface of a container and a surface is two-dimensional. He does not regard Aristotle as accepting this common conception. (J.O.U.)

[56] Reading *holôs* instead of *holos* at 620,4. (J.O.U.)

[57] Reading *sômata* instead of *asômata* at 620,5. Cf. 619,34. (J.O.U.)

[58] The intelligible gap, *khaos*, would be the abyss (above 614,5, n. 38). Place as void, Aristotle *Phys* 4.1, 208b29; Plotinus *Enn.* 6.8.11; Stoics and derivation from *kheô*, *SVF* 1.29 (Zeno). *khôra* = space, Plato *Tim.* 52B.

body present save that which is in motion and the medium through which it moves. But the interval does not belong to the body in motion (for if it moved in its own interval it would not be said to exchange one place for another), nor does it belong to any other material body that is the medium. For it is not through the air or water in a static condition that that moves which is said to move, but when it is cut and moved aside. The interval through which moving objects proceed is, then, that which is left vacant after the cut, which receives the things that make their way through it. The size of the interval covered by the motion is measured as so many times that of the moving body, depending on how many times bigger it is. Thus there is not only the interval belonging to the bodies, but another independent of the bodies.

Aristotle's argument maintaining that no other interval is left save that of the body itself is sufficiently obscure, beginning at 'If there were an interval naturally remaining in the same location' [*Physics* 211b19] and ending 'which is the place of the whole heavens' [211b29]. It has already been said that something connected with this follows from the saying that if there is an interval naturally remaining in the same location there will be an unlimited number of places, and place will change its place, and there will be a place of place, and there will be many places coinciding. But these things are impossible; so it is impossible that there should be such an interval.

This whole argument starts from the basis that the parts of a thing in a place are also themselves essentially in a place, if an interval pervades each of the parts. If that is so, then, if the thing in the place is divided without limit, the result is that there will be places without limit; and if the supposed vessel is moved, the places will also move into other intervals. Thus place will change its place, and place will be in a place. But these consequences do not follow for those who say that place is the limit of the container because the parts of the things in the place will not be essentially in a place.

That these consequences do not flow from Aristotle's argument has been shown in the discussion[59] of that passage.

[59] The discussion in Simplicius' commentary on the early part of Book 4 of the *Physics* that precedes the Corollary on Place. (J.O.U.)

For as the body is continuous, so those who hold this position suppose that the interval is also continuous and immobile. So, just as the parts of the body are not as such divided, since if they were they would not be parts but rather bits and individual things, and just as they are not completely unified either, for that is the nature of elements, but their whole existence is one of being separated, so in the case of place. As Aristotle himself was accustomed to say, the parts of both body and place exist potentially and not in actuality.[60] Since the parts of a body have that character, they are as such essentially in a place having that character and not contingently. If this is true, not everything that is essentially in place needs to be divided, as Aristotle says. If it followed from this that it must be limited by something, that also is not altogether necessary.

But nor will a place exchange its place. For the whole interval is immobile, receiving in each of its parts whatever bodies continually come to be in them, just as water, running through aqueducts that are immobile, is contained at different times in different parts of them, and does not drag the previous part of the aqueduct along with it. So, also, place will not need a place, nor will there be many places coinciding. For the intervals of place do not exchange their place nor make their way into each other. However, there will not be places without limit, if neither the parts of a continuous body are actually divided into innumerable parts, nor those of the interval. So I think that those who say that place is an interval can in this way easily escape the absurdity alleged by Aristotle.

But let us consider the other problems that may have a bearing on this view. We must first enquire whether the interval underlying at any time is material or immaterial in form. For if it is immaterial, then place, the subject of our present enquiry, will not be part of nature. Again, if it exists only notionally, then the thing in the place will exist only notionally, and not actually, and, if this is so, the mathematical object will be a body. But if this also exists actually, how will something wholly immaterial be sublunary,

[60] Parts of body and place exist potentially not actually, Aristotle *Phys.* 4.5, 212b4-7.

622,1 three-dimensional and an essential, not contingent, being? But if someone were to say that it is material, how would the three-dimensional form endowed with matter differ from body? For if we were to say that it is merely dimensions, it would be very different from that which has the dimensions. For it would be merely a form, the other enmattered form.[61] But if we say that the interval is also a materialised form, and say that the composite body is dimensional, still, how will it differ from body? Again, if it is a materialised form, it either takes the whole of matter or part of it. But if the interval is a part and the body is a part, then either two matters will coincide, that of the place and that of the thing in the place, in the same interval, or the place will be somewhere and the body in a place somewhere else, so the body is not in a place, and the place has no body. But, if the interval took on matter completely, how, having received it, would it also receive body throughout itself?

Secondly, if place is one thing and that which is in place another, either there is a single matter, but two forms, or there must be two matters, in addition to the forms. If there is a single matter, when that which is in a place leaves it, its matter will depart and it will be separated from itself, remaining because of the place, exchanging its position because of the body. But if there are two matters, how can there be two in the same volume, that of place and that of body?

Thirdly, if the interval of body comes to be in the interval of place, how is it that we object to there being two bodies coinciding, but agree to there being two intervals? For other characteristics of body, such as whiteness and heat, do not prevent interpenetration.

Fourthly, if body is a substance in a place and place is an interval, and an interval is a quantity and a quantity is a characteristic, a substance will be in a characteristic. In itself this is absurd, and, moreover, we say that place is superior to that which it contains.

[61] Proclus distinguished the (Platonic) Form, which is separable from matter, from the inseparable (Aristotelian) form, which was surnamed 'enmattered', *enulon eidos*. See *in Parm.* 863ff. The vicious infinite regress known as the Third Man, Plato *Parmenides* 131A8-132B2, is avoided for separable Forms by postulating that only participation in immanent, material forms involves a physical sharing of property.

Fifthly, this interval is either animate or inanimate. If it is inanimate it is also without intelligence; for intelligence cannot be present in anything inanimate. And how will it not be inferior to what is in place? For intelligence is in the soul, but the god built the universe with soul situated in body, as the *Timaeus*[62] says. But if it is animate, how will it be changeless, having a self-changing soul? Or, if it does change, will it not need a place.

Sixthly, is the complete interval perpetual both as a whole and in its parts, or is that part which received the heavens perpetual, but that which is sublunar perishable in its parts? If the complete interval is imperishable, how will the sublunar differ from that in the heavens? But, if it is perishable, what will be the way in which it perishes, since the parts of place will not exchange position with each other, nor will place turn into anything else, nor anything else into place?

Seventhly, if things not in need of a place are indivisible, then the divisible will need a place, as a logician might say.

It is possible to raise these and similar objections against 623,1
those who say that place is an interval. But one should not give in altogether to all the questions raised. For we shall say that the interval is material in reply to him who asks whether it is material or immaterial. For just as we say that that which is fragmented and has a volume and is indefinite is material, but that the boundary, the limit and the shape are formal, we shall say the same about intervals. For here, too, we shall say that their extensions, which both receive things and give them up, are material, but that the measured and the bounded, or rather the measure and boundary of the intervals, is formal. This form clearly came to be in a portion of matter, just as did the form of the heavens. For there is one type of alteration in regard to spatial form and another in regard to bodily form. Also there is nothing absurd in two matters being coincident. For matter does not prevent mutual penetration. Nor is there anything absurd in two intervals coinciding if one is corporeal, the other empty, and one is a kind of space, the other in the space. For 'interval' seems to

[62] *Timaeus* 30B, god = Demiurge. (J.O.U.)

have four senses: (1) only in an unextended account, such as the definition of an interval, (2) in the conception of extension, as in mathematics, (3) of a material thing with natural qualities and resistances, such as a body, (4) of a material thing, but one wholly without qualities and incorporeal. Material separation is something in addition to these, viewed as the fragmented and indefinite.[63] Nor is place a characteristic but is itself a substance. For it is not simply extension, but an extended space. Nor is there anything wonderful about it being both animate and immobile; for psychic life is not the cause only of change of place, but also of being stationary. Thus many animate things are static, such as the celestial axis and the poles and the centre and the whole earth. If the soul is said to be the initiator of change, it is with reference to change viewed as including immobility, which is more accurately called coming to be. There is nothing wonderful about the whole interval being perpetual while the heavenly differs from the sublunar. For all the heavens, while perpetual, differ greatly from each other. 'But', they say, 'if those things that have no need of place are unextended, the extended will be in need of place'; but they ought to have added the words 'or is place'. It is the same as if one were to say that if things that are unchanging stand in no need of time, things that change are either in time or time itself.

But may I say jointly to those who say that place is a body and those who say that it is an interval that if it extends right to the centre then not only what are called immaterial firmaments by those who do not know Greek[64] but also all the celestial bodies of higher space, extending ever through that which is below, will be in the same portion of the interval of place, the empyrean body together with the aetherial and those of each of the heavens, and each of the generative elements,[65] of fire or

[63] Proclus exploited the different senses of *aoristia* and *apeiria* to produce a scale of infinitude ranging from boundless divine and intelligible potency to infinite corporeal divisibility and indefinite material potentiality. See E.R. Dodds, *Proclus, the Elements of Theology*, 246-8 (on *El. Th.* propositions 89-92).

[64] *stereôma* means solid = firm-ament. The 'Barbarians' in late antique astronomy and Neoplatonism often are the Chaldaeans, Proclus *in Tim.* 3.125,27-126,5; 3.249,12-250,9, or synonymously the Assyrians, Simplicius *in Phys.* 643,27. The fame of Babylonian astronomy and celestial religion was widespread. For a study, O. Neugebauer, *The Exact Sciences in Antiquity*, New York 1969, 97-177.

[65] *genesiourgôn stoikheiôn*: Neoplatonic technical term for elements involved in the

air, or water, or earth. If then we say that a place properly belongs to things in a place, to which of all these bodies shall 624,1 we say that place properly belongs? We must also say both to these theorists and to those who say that place is the limit of the container that they are giving an account of the explanation of the position of each of the perpetual parts in relation to the whole. For why is the earth in the middle? Because that part of the body or interval of place or the limit of the container is proper to it. Why is one part of the fixed heaven the north, another the south? Or of the planets, why is one outside, the other within the spheres? Or, in my body, why is the head above, the feet below? And why is this hand and perhaps my liver on the right, but that hand and my spleen on the left? However, if these are differences in places, place should certainly be the cause of them also. But how could one of the above-mentioned places, whether it be a body or an interval or a limit, be an explanation of these facts? For my head occupies one portion of that interval when I am standing up and another when I am reclining or sitting, and one when I am here, another when I am somewhere else. But it always retains the same position in my body. Also, how will the liver always be in the parts on the right? For of what is this the right? Not, presumably, of the interval or of the limit.

This, also, must be said to all of them jointly, that, if the function of place be not well defined, one cannot discover whether place is a body or incorporeal, nor whether it is divisible or indivisible in its nature. We must ask those who say that place is a separate interval or a body whether they say that the parts of the heavens are in place, or not. For if they are not in place it is clear that they will not change their place; for what changes its place must be in place. But, if those portions do not change their place, what other sort of alteration do they undergo? If they change their place and are in place, whether it be body or interval, it will either be in their proper place or elsewhere. But, if they are in their proper place, each will remain in it and not move; for what need have things in their proper place to move, when they are going to be unnaturally positioned if they leave their natural

production of becoming and perishing, Iamblichus *de Mysteriis* (Parthey) 1.11; Proclus *El. Th.* proposition 209.

place? But if each is elsewhere than in its proper place, what sort of body will it be that is always so and never in its proper place? Someone might say that perhaps the whole interval of the zodiac or the body constituting place is proper to everything in the zodiac and that therefore each part is and moves within the whole of it; but first it would also have to be at rest in the whole of it, which is impossible, and also what of the postulate that says that a place must be equal to that which is in it? One could raise the same problems regarding the parts in the four elements.

Since, in addition to those discussed above, Damascius, the philosopher from Damascus, a man devoted to research who introduced many philosophical labours, constructed another theory, we should expound it also. We shall of necessity need a longer discussion on account of the novelty of his theory, which seeks to discover the nature of place from the purpose it serves.

His statement is as follows: 'Everything in becoming that fell away from the unified and unextended nature in essence and in activity suffered a double separation both in its essence and in its activity or passivity. That in its activity was also twofold: one naturally linked to its being, through which its being is in continuous flux, the other proceeding from its being, through which it acts in different ways at different times, since its activities are diachronic and not synchronic.[66] The division of its activity immediately demanded change, and change came into being with it, and its division came about through change in activity or passivity. The division in its essence was also twofold, one being fragmentation into plurality, the other the acquisition of bulk. The division involving size and bulk in a position came about through the scattering of the parts in various locations. But this, too, had a double nature, one linked to the thing's being, such as in my body the head being above and the feet below, the other external, such as my being positioned sometimes at home, sometimes in the marketplace. Of these, the first clearly continues the same so long as the thing lasts, the other is different on different occasions.

[66] i.e. activities act in an orderly sequence of moments (*paratetagmenas*), not all at once (*athroas*).

Those things are properly said to have a posture[67] that have adjoining and extended parts. On this account position seems properly to belong to magnitudes and their limits, because they are continuously extended. At any rate numbers, although they are separated from each other, do not seem to have a position, because they are not extended and adjoining, unless they also acquire size and extension.[68] For all intervals, having lost their unified wholeness, exchange their being in themselves for becoming in something else, which includes posture, having, as it were, been set aside and having lost their self-government. Similarly in activities things that lose their nature are said to change and suffer alteration. In order that these extensions should not be produced indefinitely, uniting measures came into existence. Time became the measure of the extension of the activity in change; of extensions in being, the determinate quantity which is number became the measure of divided extension, the determinate magnitude, such as the yard, became the measure of the continuous, and place of the scattering in position. Therefore changing things are said to change in time, and to have their position and its changes in so far as the position itself changes, in place.'[69]

Thus it is very clear that place is concerned with position and with things having a posture. For we say that those things are in a place which we also say are in a position; and above and below are differences of place, viewed in relation to position, just as are right and left and in front and behind.

That place marks off, measures and orders position can be learnt as follows: we say that each thing has a posture,

[67] The Greek word is *keisthai*, which is the name of an Aristotelian category. (J.O.U.)

[68] Numbers acquiring size and extension refers to the transition from arithmetic to geometry and stereometry. It is Neopythagorean in origin. The Neopythagorean influence from the second century AD is great. The Neoplatonists taught the Neopythagorean arithmetic of Nicomachus as part of the Quadrivium of Mathematics: arithmetic (numbers), geometry and stereometry (dimensional objects), astronomy (dimensional objects in motion), music (harmonics of interval). 'Acquiring depth', Sextus Empiricus *Adversus Mathematicos* 10.281,4-282. D. O'Meara, *Pythagoras Revived: mathematics and philosophy in late antiquity*, Oxford 1989.

[69] The close comparison of place and time is a noteworthy achievement in the history of science; cf. also 632,33 'place resembles time'. See S. Sambursky, *The Concept of Place in Late Neoplatonism*, Jerusalem 1982, 15 (notes, text and trans. (incl. of the Corollary)).

however disordered this may be, but a thing is said to be properly positioned when it gains its proper place, just as a thing comes to be whenever it reaches being, but is opportune when it comes at the right time. So, because place properly positions each part, the head is at the top of my body and my feet are at the bottom, and my liver is to the right and my heart central. Also the eyes, through which we see as we advance, are in front, but the shoulders with which we carry burdens are behind. These differences are because of place, just as because of time the parts of the embryo are framed one after the other, and one age precedes another in proper order. The Trojan war is not mixed up with the Peloponnesian war, for before and after are differences of time, just as up and down and the other four directions[70] are differences of place, as Aristotle confirms. So, therefore, the parts of the universe have their proper positions in it because of place.

So, to speak in outline, place, according to this view, is, as such, the demarcation of the position of bodies, but, to speak of place in nature, it is the demarcation of the position assigned to the parts of bodies in relation to each other and to the whole, and of the whole to the part. For, as the different parts of the earth and the heavens are differently disposed on account of place, some being perhaps to the north, others to the south, so also the whole heaven and the whole earth are parts of the universe and have their proper measure and arrangement of position because of place, one retaining the outside of the whole, the other the middle. Again, it is place that brings about the coherence (*sumptôsis*), that is, the coplacement (*suntopôsis*) of parts. But if 'place' (*topos*) is derived from 'divine' (*topazô*),[71] place being so-called from a divination of the adjacent, as being a kind of imaging of the intelligible distinction, that is also a consistent account; for place provides a structure and likeness to the paradigm for

[70] These are front, behind, left and right. (J.O.U.)

[71] Concurrence (*sumptôsis*) is the combination of the appropriate constituents forming the whole; coplacement (*suntopôsis*) is the fitting location of the said parts; *topazô* means to guess: S. evidently wants to link semantically the putting into place by comparison or conjecture (guessing) with the act of imaging (*eikasia*) (in Greek both mean to guess). Imaging, and imitation, refer to the Platonic relation of sensibles to Ideas as exemplars. The meaning of this difficult passage is, therefore, that place (*topos*) arranges and locates things by imitating and copying the formal distinctions in the intelligible.

the images. For if each part were not situated in its proper place when they had once been divided, the image would never have resembled the paradigm, but every order, good measure and arrangement would be gone. If you were to abolish place, you would see the disposition of bodies as uncouth, disarranged and moving into indeterminacy. For in what position will each of the parts take its station if it has no affinity to any? That is why moving things naturally move, in order to obtain the advantage of their proper position, and stationary things stand in their fitting portion of position, that they love. So from what has been said one can easily come to see what place effects for bodies and everything corporeal, and what it is itself.

It follows that place, if it is of such a character, is not the limit of a container. For how is that the cause of order, or distinction, being itself rather marked off by the things which come into and are contained by it? But it could not be a body. For even if one said that it was an immaterial body, having itself separate portions and differences in the portions, it would need itself something to put it in order and make one part be in the middle and another at the outside. Nor can such a thing be an interval; for the interval, for the same reasons, containing differences and different parts differently disposed, will itself need to be set in order.

So place seems to be the measure of the position of things in a position, just as time is said[72] to be the number of the change of changing things. But since position is of two sorts, one essential, the other externally imposed, so place will be of two sorts, one being an element perfecting that which has a position, the other contingent. There is a further difference within essential position, in so far as either the wholes themselves contain the proper position of their proper parts in relation to each other and to the whole, or else as parts have it in relation to the whole and to other parts. In this way place comes to be of two kinds, one suited to each of the parts individually, the other distinguished according to the thing's position in the whole. For just as the whole is of two kinds,[73]

[72] By Aristotle at *Physics*, 219b1. (J.O.U.)

[73] Kinds of whole: cf. Proclus *El. Th.* propositions 67-9, who distinguishes a whole that is not analysed into parts, a whole as the sum of parts, and a whole as the

one suited to each of the parts in accordance with its distinct and separate existence (in this way the earth is said to be a whole, but not only the earth, but also each animal and plant, and their parts), the other more widely embracing, as we say that the universe is a whole, the earth a whole and the air a whole, in the same way we say that place of one kind is the well-positioning of the parts of each thing, as the parts of my body in the whole; of the other kind is the well-positioning of the whole body as a part in the place of the more widely embracing whole. In this way the place of dry lands is the place of the earth, and in this way the earth occupies the middle of the whole. For if one removes the earth from its position around the middle of the whole, it will retain the well-positioning of its proper parts in its whole self, but will no longer have it as a part of the whole. Therefore, if the whole earth were set free it would bear towards the middle;[74] yet its parts would retain their relative disposition, even if it itself happened to be outside the middle. Similarly, a man up in the air will retain the good order of his own limbs, but has lost his own as a part in relation to the whole.

Parts belong to greater wholes more basically than they are themselves wholes.[75] For they are governed by superior beings to a greater extent than they govern inferior ones, because the first is always proportionately greater in relation to the second than is the second in relation to the third. So the clod of earth may retain its interior good order, but it will justly move, striving towards the greater whole. For the individual is dead and lifeless when it is torn away from the community and deprived of its proper connection with it, just as plants, if pulled out by the roots and retaining all their own parts, will immediately wither when torn from the common wholeness. For everything lives through the living cosmic One. Thus, so long as each thing is rooted in the universe via its neighbouring wholenesses, that long it lives and is preserved; but, if it is torn away from the neighbouring whole, it is torn from the common whole.

---

individual part existing distinctly. See also A.C. Lloyd, *The Anatomy of Neoplatonism*, 76-85.

74 Silently contradicting Aristotle *de Caelo* 294a16ff. (J.O.U.)

75 This is the standard Neoplatonic hierarchy of wholes and universals; see Proclus *El. Th.* propositions 70-2.

In this way, then, the natural motions of bodies and their rest in their proper places are preserved by him who posits a place of this sort. Change of place by moving things is nothing other than the exchange of positions, one after another, until the moving body attains its appropriate position, penetrating the air or water in between, which retains the position it then has until a more powerful body comes along. There is a position of the parts of the air which the clod or I take over as we move. This is place in the wide sense that I take over, not my own private and limited place, but that of the surrounding air, of which I naturally occupy different parts at different times.

Since it is problematic how things that change their place move, if it can be justly said that things in their place are at rest rather than in motion, let us see how the philosopher Syrianus presents the problem and offers his solution to it. He writes as follows: 'How, some might ask, do things in motion move in place? Things in motion are, more accurately, in transit from one place to another; for, in general, things in a place seem to be at rest. So perhaps things in motion are both in a place and not in a place. For they are not in their primary and, so to speak, private place, since, if they were, they would be at rest; but they are in their more broadly conceived place, as when we say that the sun is in Leo, since the area of Leo contains it, and we say that the eagle is in the air and the ship running before the wind is in the sea. Now all these have a place, broadly conceived, but they are not in their own primary place so long as they move.'

It seems to me that the majority of those who discuss place think chiefly of this exterior place. For, if asked what is the place of the earth, they reply that it is the middle of the whole, as if the place of the whole were private to the earth as being within the whole. If asked the place of the heavens, they say 'outside'. But they do not take into account[76] the place of the earth itself, that sets its parts in order, and the similar place of the heavens, and likewise with particular animals and plants. This, it seems, is the reason why all have represented their place as being separate from the occupants. For, in

[76] Reading *proslogizontai* instead of *prologizontai* at 629,4 as Diels in apparatus. (J.O.U.)

reality, the more comprehensive place allotted to each thing is separate from the actual thing in that place. Also they claim that place is immobile with an eye to this common and broadly conceived place. For the proper place of each thing that essentially belongs to it certainly moves with it, while the common broadly conceived place is at rest and is proper to a more comprehensive and more inclusive body.[77]

But let us also, if we may, look at the objections to this account.[78] First, having proved that place is not the form (*eidos*), we now say that it is the well-ordering of the parts, even though each species (*eidos*)[79] has its own ordering of parts, according to which it is what it is, as, for example, man is upright, but the irrational animals face towards the earth, and some have their parts arranged in this way and some in that. How then are we to say that it is not the specific form (*eidos*) that is the cause of the ordering, but place?

Secondly, if we say that place is the well-ordering of the parts, something that is not a part could not be said to be in a place. How, then, does this account not seem to have the same flaw as that of Aristotle, which does not admit that the whole universe is in a place, if it is not the part of anything nor cut off from some other part set against it?

Thirdly, are we to say that the moving parts of the heavens change their place, or not? If not, how shall we not be entangled in the same contradictions as was Aristotle? But, if they do change their place, what is this place? For what is changing its place ought to vacate one place and take over another. Nor is the place of its parts, through which we say that the parts are coordinated, of that kind, for that remains ever the same; nor is that which coordinates the other parts according to their kind, through which the Bull is placed after the Ram and after the Bull the Heavenly Twins, and so on. For these arrangements remain always unchanged. But,

[77] The comparative cases, 'more whole' (*holikôterou*), 'more comprehensive' (*periektikôterou*) illustrate the Neoplatonic hierarchy: the more whole cause contains more effects as parts. See Proclus *El. Th.* proposition 177; Damascius *Principiis* (Ruelle) 1.200,20-5; 1.238,23-9; 2.179,1-5 (= *in Parm.*)

[78] The following ten queries emulate Damascius' Ten Questions and Answers in *Principiis* (on the nature of the remaining, procession and return; Westerink, 2.139-73 = Ruelle, 1.175-95).

[79] The Greek *eidos* covers both form and species, since the Greeks conceived of a species intentionally, not extensionally. (J.O.U.)

further, the view of place that is now presented does not take into account a more comprehensive and unchanging place, within which parts are said to change their place by exchanging their position. But how, if the disposition of the whole is always altering in respect of the relative position of the parts, will the place of the whole also not alter, so that there will be as many places as there are dispositions?

Fourthly, how will the place of a thing in that place be superior to it, if it is an element in the thing? For the element is inferior to what contains the element, as the part is inferior to the whole.

Fifthly, place is either extended or it is unextended. If it is extended, how will it not itself need a place, as has been said? But, if it is unextended, how can it be equal to or contain or be receptive of things in place? For what is equal to an extended thing ought to be extended, and what contains or receives a thing with volume ought itself to have a volume.

Sixthly, how shall we not appear to say that place is inseparable from what is in place, when almost everybody believes that it has a nature separate from that of the thing in place?

Seventhly, how will it be unmoving if it is an element of the thing in place? For it will move together with it.

Eighthly, if we say that place is an element of what is in place, why do we not rather say that it is in the body? But body is in place.[80] For an element is in what contains elements, not the other way round.

Ninthly, if place is the measure of position, how can what is accepted by most philosophers be true, that the perpetual signs of the gods in the universe are at rest because they are inscribed in the place that is perpetual, and by these means bodies in place also enjoy the divine illuminations?[81]

Tenth and finally, if we arrived at the idea of place from the exchange of position by bodies, since where there formerly was water there came to be air, how, if everything has its own

[80] Question mark instead of comma after *einai* and full stop instead of question mark after *topôi* at 630,10-11. (J.O.U.)

[81] The 'parts of heaven' = (zodiacal) constellations; 'the gods *in* the universe', the encosmic gods = the celestial bodies; the 'divine illuminations' = the celestial, astrological influences (note the term 'influenza' in medieval medicine, for the unearthly cause of the illness).

proper place, shall we have need of this exchange of position? These and similar points could be raised in an attack against this apparently intellective conception of place.

In reply to the first objection, it should be said that the whole form is a manifold of all the elements, containing each in relation to each. Just as it is a whole not on account of the beauty that it may chance to possess but because of being whole, and complete on account of its completeness, and similarly for its other elements, so it has an ordered and well-shaped position because of its place. We should not separate its form from its participation in its place that jointly contributes to its being, and wish to see it well-disposed entirely in itself. That is as if, having conceptually taken away its completeness, we should strive to see it as complete even without its completeness. So this account of place is in no way contrary to one that denies that place is form, even if it holds that place is an aspect of form. If the proof that the form is not place used as a postulate that place is separate from what is in place, and, making use of another premiss saying 'the form of what is in place is not separated from it', concluded that place is not the form of what is in place, one must recognise that this postulate is introduced in accordance with certain conceptions of place with which the sense of the account of space presented by our colleague Damascius is not compatible.

631,1 Also, even if the notion of place is taken from the exchange of place, by which we see air coming to be where perhaps water was, in this way we find place as broadly conceived and not that place proper to each of the parts. For the proper place, because of which the same place[82] belongs to the same body, remains unchanged, but the place which is occupied by different things at different times is consequently not a proper place.

In reply to the second difficulty raised, it should be said that, even if we say that place is the well-ordering of the parts, none the less the whole, also, is given its situation by it. For just as what is well-tuned, even if it is well-tuned in virtue of the proper proportion of the parts to each other, is still said to

[82] Reading *topos* instead of *logos* at 631,5. (J.O.U.)

be well-tuned as a whole,[83] so what is arranged and well-ordered, even if it becomes so in virtue of its parts, since all ordering is through discrimination, is still itself what is arranged. A thing put into its place will be in the same case. Also a measured whole is measured through its parts. But we say that place is a certain arrangement and measure or demarcation of position. When we say that the whole is unified, do we not say that the whole is unified through the unification of its elements? Yet, viewed as a whole and one without distinction of parts, we should not say that it was harmonised or unified, but that the whole is unified with respect to its parts. So we say that a body is well-positioned and well-placed just as having each of its parts in the appropriate position.

You must see that just as we say that the parts belong to the whole, so we say that the places of the parts in which it is situated are parts of the whole, such as the middle and the outside; reasonably so, since there are parts of the whole position, just as low and high notes are part of the same scale. So there is, in truth, the whole place of the whole universe, but it has its supreme position through the good arrangement of its parts and through its whole good arrangement in respect of its parts. For differences of place, the above and below, belong to the whole and are judged in relation to it, just as, if my head alone of my body happened to exist, it would not be said to be situated above according to the above in the total being. In general, we do not say only that the parts have a well-arranged position in relation to each other and to the whole, but also that the whole has it in relation to its parts.

In reply to the third objection, it occurs to me to say, as has been said before, that the position of the whole and of each of its parts is of two kinds, the one essential and unchanging, the other contingent and varying from time to time. Within the essential, there is the position of the whole as such, and its position in relation to its parts, and again, the position of the parts as such and their position in relation to each other and

[83] S. uses the example of a musical instrument, the lyre or the pipes, to show that the whole is well-tuned – in harmony – when its parts are in their right proportion – in symmetry with each other. The late Neoplatonists taught music (mainly from Aristoxenus, fourth century BC) as part of the Quadrivium of Mathematics.

to the whole. Now the whole essential extension or position always remains the same, whether the parts be viewed as moving or as stationary. The place that unifies and measures
632,1 this extension is ever the same, but the varying place is altered. For not only do the parts of the heavens exchange their position, but the whole also.[84] For the position of the whole heavens now is different from that an hour later. For this reason both the parts and the whole could legitimately be said to change their place, since they alter the order and demarcation of their position, which is what we said that place was. Just as we say that those things change in time that leave one time and take over another, even if the first time comes to an end, so those things which leave the demarcation of their position and take over another may be said to change their place, even if the first place comes to an end. In the same way we say that things exchange their quality if they were formerly white and later become black, not because the whiteness that the blackening thing discards remains behind on its own. Also things that come to be and pass away alter, as do things that quantitatively increase and diminish, though their previous condition does not remain behind. So we should not require former places to be preserved when things exchange their place. Nor in our case when we change our position are the places preserved, as when I cross from here to there; but it is the whole of the surroundings that are preserved, and their place, which is able once again in one of its parts to coincide with my interval. Just as I, though in a different position, retain potentially a coincidence with that part of the whole and the demarcation of its position. For this reason I am able to again become coincident with[85] it and to have the demarcation of my position, i.e. my place, in that part, since the whole place of the air measures my extension in a portion of itself and gives me a position in the whole of the air. Also, as Aristotle himself presupposes, the necessity of vacating a place does not mean that the boundary of the air containing me remains when I withdraw from it.

[84] S. attempts to solve the difficult problem of the changeability of the heavens as a motion relative to the place of the whole. Cf. also S. Sambursky, op. cit., 152, n. 34.

[85] Become coincident with, adjust to (*sunarmozesthai*), also to fit in, to adapt to, to

But if we say that it is the parts that exchange place for place, because the place of the whole remains immobile, that, too, I think, will be preserved on the present hypothesis. For the essential demarcation of the position of the whole is always at rest, whether the total universe moves or is stationary. For, if the whole is at rest, the multitude of varied succeeding positions in the universe is a sort of unfolding of it.[86] For the one essential position embraces every position of the total universe. The same holds for each of its parts.

In that respect, place resembles time. For one sort of time is perpetual and remains in the same state, being of like nature with the coming into being of the substance of the heavens which are in a way ungenerated, and having only a suggestion of becoming in respect of bodily extension and alteration of all kinds – if, indeed, one is prepared to call this and what measures the being of the soul time, rather than something between time and eternity. The other sort of time has its whole being in coming to be and passing away, and measures the changing movement of the heavens and all kinds of alteration in the world of becoming. In the same way one sort of place measures extension or position in the world of being and is perpetual and static like that extension, while the other sort of place measures what alters and is seen as active, and has its being in becoming.[87]

Just as we say that the continually occurring natural emanations of the life of the soul are an unfolding of its life that pre-exists as having being and has stored within itself according to fixed rules the separation out of the ema-

---

combine with, and in music to tune in, *lura sunêrmosmenê pros tên aulon*; see *LSJ* s.v. *sunarmozô* 2.

[86] 'Unfolding' (*anelixis*) is another Neoplatonic ref. to the procession (see nn. 1 and 33) from one to many, applied especially to the 'unfolding' of parts from a whole. Cf. Damascius *Principiis* 2.52,21-53,19, Westerink (= 1.120,3-18, Ruelle). In this passage, the unmoved place is the essential place that 'remains' unified and delimited. The moving place is that which 'proceeds' as the place of a part, the place of each particular body.

[87] Levels of time and place are characteristically Neoplatonic, and are part of the ontology of participation and the triadic scheme 'remaining' (*monê*) – 'procession' (*proodos*)-'return' (*epistrophê*). The unmoved, perpetual time is the measure that 'remains' above the cycle of becoming and perishing, but the time of change is the measure that 'proceeds' in temporality. Iamblichus and Proclus distinguished eternity proper, which is at the level of intelligible being, from unmoved time, which is at the level of intellect itself (following the distinction of the intellect from its object).

nations;[88] similarly we say that the essential place of the universe has stored up all the varying places and produces from within itself the proper measure of every position. The position of the Ram in the world of being has predetermined all its various positions, and its place all the places. For the Ram does not have being because of its rising or of its setting or, in general, because of this or that position within the whole – just as the soul does not have its being through some individual emanation – but through its unitary anticipation of every position of every kind.

So, even if place were not presupposed as an unchanging body or interval, the heavens are not prevented from changing place in any way. For men also say that they change in time without presupposing some changeless sphere of time whose parts will be said to alter as they change in time. But, if there is a need of something changeless, they are happy with the time in the world of being or they have recourse to the paradigmatic time.[89] But perhaps the place which is supposed to be unchanged contributes very little to change of place. For, consider, if that which conceptually is the sphere of the interval or of the corporeal place were taken away, what else would we call the motion other than change of place if the heavens moved with their circular motion? But that changeless sphere seems to help our judgment, so that, by comparison with it, we may estimate and calculate alterations, just as astronomers have supposed the starless sphere[90] and its intelligible zodiac, and the intelligible periods

[88] Emanations, projections (*probolas*) of soul: Neoplatonic ref. to distinct reason-principles (*logoi*). In an important passage Proclus *in Tim.* 2.123-4 explains that entities involved with generation cannot receive and contain the infinite power of intelligible being – which he links to Aristotle's statement in *Phys.* 8.10, 266a25, that unlimited power cannot reside in a finite body. Soul cannot contain at once the whole of the intelligibles, because it is a created, not an eternal entity, being a creation of the Demiurge, which is Intellect (Proclus calls also on Plato *Tim.* 35A, the position of soul between the intelligible and body, and to 37A, soul as a generated thing). Thus soul actualises its intelligible contents by 'projecting' them in time and place (cf. Pr. *El. Th.* proposition 191). Further, in *in Euclid A* (Friedlein) 13,10; 51-6, Proclus conceives mathematical objects as projections of soul.

[89] Paradigmatic time refers to Iamblichus' intelligible, unmoved time: see the Corollary on Time; also, S. *in Cat.* 352,14; 353,19-356,7; Proclus *in Tim.* 3.31-2; 33. See Introduction.

[90] The 'starless sphere' is the sphere of the precessional movement of the fixed stars, see *in Phys.* 644,1, *in Cael.* 462. Precession as a full 'apokatastatic' backward movement should be distinguished from the periodic variation in the sidereal position of the vernal point (equinox) of limited amplitude, known as the 'trepidation'. The

of the stars, not in order that the heavens might move in their circular course on account of them, but so that they could calculate the measures of their motion by comparison with fixed limits. Comparison with the unchanging also helps us with regard to their apparent change of place. For how could we say that something which does not change its position changes its place? How would we perceive that the Ram had changed its position if we did not compare it with the cardinal points of the ecliptic? But we define these in relation to the earth. If, then, there were no rising, nor setting, nor stellar 634,1
culmination, still the heavens would move, but we would have no way of judging the different positions.

In reply to the fourth problem, it should be said that place so defined, as a source of perfection, is in the same case as other perfections, such as beauty, goodness, wholeness and the like. Just as each of these seems inferior to the manifold that includes it, but superior when contrasted with the remainder, so is the case with place. Compared with the well-ordered object, the well-ordering that arranges the whole seems less complete, but as giving perfection it seems superior to what is perfected.

In reply to the fifth of the problems, which enquires whether such a place is extended or unextended, we say that place is a prerequisite as a measure of extension in position. Every measure requires to be equalised to what is measured, or, rather, to make the object measured equal to itself. But the transcendent measure, that exceeds the object measured through its superiority in power[91] and its unity of being, carries up the measured object towards its own concentration,

precession was first recorded by Hipparchus and rediscovered by Ptolemy. Proclus rejected it (L. Siorvanes, 'Proclus on the Elements and the celestial bodies', PhD diss. London 1987, 261-70). However, on S.'s evidence his teacher Ammonius confirmed it. S. proposes that the fixed stars, which normally move diurnally, are moved backwards along the ecliptic by an additional sphere, which lies beyond them: thus it is 'starless'. Later in medieval cosmology, this extra sphere was assimilated in the multitude of spheres both physical and metaphysical, derived from the empyrean and aqueous 'spheres' in Genesis and the Neoplatonic-inspired Celestial Hierarchy of pseudo-Dionysius the Areopagite. For studies see O. Neugebauer, *A History of Mathematical Astronomy*, Berlin and New York 1975, 296-8, 631-4, 909-13; E. Grant, 'Cosmology', in D.C. Linberg, ed., *Science in the Middle Ages*, Chicago 1978, 265-302.

[91] Superiority and concentration of power and transcendency: Neoplatonic expressions for the higher echelons of the intelligibles: the closer to the One, the more separable, really real and potent existence is, and the less diversified and particulated.

giving a share in itself so far as is possible, just as we say that the intelligible numbers measure everything through their concentrated supremacy over all things. Such entities are measurers rather than measures.

The well-ordered condition, as being a participation by the measured in the measurer and being coordinated with the measured object, is extended with and stretched out beside it, just as our yard is an extended measure deriving from the unextended measurer, that is, the yard in the soul. So also time is extended together with change. There is the same problem with time also, as to whether it is itself extended or unextended. It is clear that this participating time, or rather the participation, is divided into parts together with change and accompanies it everywhere. It unifies its spread and scattered parts, or, to speak truth, it exists as the unification of the parts unified. The place also that exists together with bodies is extended with them. For everywhere it is the position of bodies and the determination of their position.

But, if place is extended, how, we ask, will it not also be in a position, and stand in need of a place – that problem that faces every theory about place? One should know that everything that participates in anything is said to participate in it as taking into itself the nature of that which is participated. For things having something from their own store do not need to import from elsewhere what they have. There is necessarily a process without limit, as, for example, extension itself, through which extended things are extended, is itself extended with them, but not in the way that things are given their character by something else and are said to be extended by participation in extension, but in the way that extension itself can be said to be extended, spread out together with what is extended, and accepts extension. So if extension would need some imported extension, we shall consider again what that would be.

If extension is unextended, how has it conformed itself to what is extended? But, if it is extended, it itself will stand in need of another extension without limit. This is why we should not postulate a change of change or a time of time or a measure of measure. Why, then, will not those who say that every place is body or a limited interval be able to escape the problem alleged against them? It was objected that if each of these parts is differently situated it must inevitably be in a place. But these objectors, too, will readily say that they hold that this body or this interval is place, but there would not be a place of place, as in fact they do say. For example, the philosopher Syrianus says: 'We shall call on those who raise these problems to look to the judgment of the older philosophers and to seek neither a change of change, nor a rest of rest, nor a place nor placelessness of place.'

If the nature of body and the measure of position had been identical, this would have been well said. But if every body, *qua* body, needs a measure, then, even if it is measured according to its extension (for extension is one thing and the measure of extension another), it is clear that extension *qua* extension, taken in isolation, does not need extension but does need a measure. But how would the measure stand in need of a measure? Rather, it seems not to need a measure, but to have extension as something imported. Place is extended through its participation in the object in place, just as the object in place is measured and located by means of place. I think that there is nothing problematic in this statement; for we say that whiteness is made corporeal because of body, just as a body is made white because of whiteness. What then is odd if participated place is extended because what participates in it is extended? Someone may indeed say that the object measured by place, wishing to lay hold on it, is spread out together with it; there would be something to say for this. But I think that it is better to say that extension as a form, that one could see as a principle among the paradigms, is one thing, extension that has been affected by a decline on account of being extended another. Therefore, though we say that the extension in the mind, pre-existing in form, or the cause of extension, through which every magnitude subsists, is first and supreme, we none the less say that it is unextended because of the indivisible concentration of intellective being; and not it alone, but everything that participates in it essentially. But we say that extended things HERE[92] are extended, not only through their participation in their special

[92] 'HERE' and 'THERE' refer to the sensible and intelligible realms respectively. The emphasis on unextended derives from Damascius; see Simplicius *in Phys.* 644,10ff.

nature but also through their decline from their indivisibility THERE. In that sense of 'extension', not only are bodies extended but also the attributes of bodies. So whiteness, 636,1 warmth and numbers and everything of that sort HERE are extended not only because of the bodies underlying them but in themselves; or, rather, they received the form of extension because of the form of the body in which they occurred, but received their generated and iconic extension[93] because of their decline from the unextended, through which body as well has become extended HERE, though, being also a body in the intellect, it was unextended THERE.

So, as the body that has position became extended through its decline, so also place that is the measure of position became extended, in the way that is possible for a measure that has declined from the unextended measurer. Thus place became extended, organising all extended things, not just bodies as people seem to say, nor extensions, if there are any such things in themselves, but also qualities and other attributes. Place organises these also, not only contingently in so far as they become extended together with body, but in their own nature in so far as they became extended themselves through their decline.[94] Further, attributes have a need of time also, not only on account of body but also in themselves through their generated being. To speak paradoxically, one might say that even a generated point itself is extended, in so far as it received a position and became the limit of a line.

But what has now been said about generated extension might reasonably become the starting point for another problem. For if the extended, though it pre-existed in the intellect as such, had no need of place because of its unextended unity, it seems HERE to need a place, not because of the specific nature of extension, but because of its generated being and its acquisition of volume and the scattering of its parts in its decline. If that is so, it is clear that place itself, that has undergone extension in its decline, will have a need

[93] 'Iconic extension', as an image, Proclus *El. Th.* propositions 65, 195.

[94] Decline (*huphesin*): important Neoplatonic term for the substantial change involved in procession, particularly the loss in unity, wholeness, causal potency, permanence, intelligibility and reality. See A.C. Lloyd, op. cit., ch. 4.

of place to organise it and time will have need of time. Further, even things distinguished by their specific natures that are lasting and extended may have needed measures conforming to their specific natures, in the way that the plurality THERE needs the number THERE and being needs the perpetual concentration of what is and magnitude THERE needs the measure of magnitude. Similarly, things that exist as images through their decline were circumscribed by declined and iconic measures; but these, being such as to unify and measure becoming, or being rather unification itself and measure itself, are superior to becoming. Having measure through their own nature, they do not need a further unification such as the other things do that are unified by them. Or, if they need it, they need a superior and less complex unification. For measures are a mean between the measure of these things and what is measured, dependent on the measurers, but rooted in the things measured.

Just as bodies, through their characteristic division, have as such no share in life, but participate in the life which is bestowed on them by the soul and which unifies them, and just as this vital participation is itself extended and apportioned throughout bodies because of the decline of the soul, but does not stand in need of yet another life but of that alone that is rooted in the soul, so it may be with place. For the nature of the measure is superior to the nature of the measured and is not in need of the same things as it is. Nor does beauty HERE, though somewhat deformed through its decline, need to participate in some other corporeal beauty, but only to be an image of the archetypal beauty. Beauty, however, is more closely allied to the other forms, while place and time have instead a superior nature, being viewed as unifying principles. Their extension is not like that of other things, seen as they are as a mean between the unextended measurer and the extended objects measured. For the measure of becoming participated in becoming to the extent that it declined from the simple form that measures becoming. But it declined in such a way as not to be wholly in becoming; for otherwise the measure would not have been a mean between the ungenerated measurer and the coming to be that is measured. Being a mean, if it itself needed a measure it would need another mean state. For the measure must be entirely a mean between the measurer and the measured, rooted in the measurer and reaching up to it, but spread out with the object measured, having one simple existence combining both. But, since people suppose place to be body in its own nature, why does not this body itself need a measure, and the interval likewise, since it is a kind of substance, having different parts differently located?

In reply to the sixth and seventh of the problems raised, we shall say that those who say that place is separated and immobile and who presuppose these as axioms about place have diverted their attention to the common place, widely conceived, as was said above [628,21ff.]. For the individual place[95] of the position of each of the parts of body is inseparable from it and moves as it moves. But the place that is widely conceived, as Syrianus also said, in relation to which change of place occurs, is proper to that more complete body, and is inseparable from and moves together with it, if it happens to move, but it is separable from a body within the complete body, that is unmoving as its place is. It is not surprising that those whose conception of place arose from the exchange of position by bodies should concentrate their gaze on place as widely conceived.

The eighth difficulty says: 'if place is an element in that which is in a place, and if the element is in the composite of elements and not the composite in the element, the result will be that we say that place is in the body and not the body in a place.' This can be resolved by saying that, if we consider the thing in place as in the more complete and more widely embracing kind of place, we can reasonably say that the object
638,1 is in the place. One can also see place[96] as being the pre-existing manifold of that which is in place, uniting its extension, in the same way as one might think of the animation of the living body. For place also has a certain transcendent character that encompasses everything in respect of place that body embraces corporeally. On that account it seems to be transcendent and separate, and body is

[95] S. contrasts the individual place (*idios*) with the more whole, universal place (*holikôterou*) (Syrianus).

[96] Reading *auton* instead of *auto* at 638,1. (J.O.U.)

said to be in place, not place in body. For the comparison treats them as separate.

So the unified is in what unifies it, and the divided in the united and the measured in the measurer, just as parts are said to be in the whole that is distinguished from its parts.[97] For the measurer embraces and contains what is measured, bounding it on all sides and unifying it. This is why it is likely that place first enjoyed the divine illumination, especially the place of more complete and perpetual things, with an eye to which they say that the signs of the gods are perpetually scattered in place, but also the partial place proper to each thing. For unification and measure and bound are more akin to the unitary modes of being of the gods and would first enjoy their illumination. So the ninth difficulty is to be resolved in that way.

The final difficulty runs[98] as follows: if it is generally accepted that the concept of place came to us from nowhere else than from the exchange of position by bodies, why, on the basis of that exchange, do we say that place is the demarcation or measure of position? For, whether bodies do or do not exchange their positions, the proper demarcation of each remains the same. This difficulty, also, is resolved by things which have already been said repeatedly. Place, as has been said, is of two sorts, one of which essentially and *per se* belongs to each body and remains the same whether the body is stationary or moves, while the other comes about through the exchanges of position in relation to the disposition of bodies in general. The essential place, being naturally joined with substance,[99] can scarcely be distinguished from substance, but the other varies at different times while the substance remains unchanged. It thus easily overwhelms and diverts our judgment by its difference from the substance. In the same way, time has a dual nature: one that measures

[97] The sum of parts is distinguishable from the whole, viz. *pan* vs. *holon*: the latter contains additionally the arrangement of the parts. See Plato *Laws* 10, 903C-904B, *Theaet.* 203E; Aristotle *Phys.* 7, 250a, *Metaph.* 1041b, 1024a.

[98] Deleting quotes at 638,18 and 23 and deleting commas before and after *phêsin* at 638,18. (J.O.U.)

[99] Neoplatonic ref. to the level of Intelligibles, or Real-Being, or Substance; derived from the Platonic theory of the really real (*ontôs on*). The permanent definition and character of a thing that can be fixed by mind is the thing's 'really real' essence. Its changeable, temporal aspect is its physical appearance.

essential change, the other external change involving activity. Of these, the time of essential change is so hard to discern, if, indeed, it is proper to call it time, that Aristotle denied altogether that things with a perpetual being were in time, admitting only those things that are measured as coming to be and passing away.[100] For essential change was unrecognised, but that involving activity is obvious.

Of positions, also, the one is essential, even if it be of parts
639,1 widely distributed. But it becomes clearer that it is something distinct from the parts when the parts are at rest in their own essential positions and the varying position of the whole can be seen. This comes about when there are exchanges of position. Perhaps the explanation of the essential position proper to each thing and its measure escaping notice was this, that it was hard to distinguish true position from what is called posture. But the more general kind of place was more obvious through changes of position, and it seems that it is this that the Peripatetics have in view when they say that only bodies that change their place are also in a place, because it was this exchanged place that they especially noticed.

Now that I have set out this concept of place so far as I am able, and now that I have set out the difficulties raised against it, and adduced the resolution of the difficulties, I wish to show that the concept is neither entirely novel nor unknown to famous philosophers. Thus, Theophrastus appears to have held this position in his *Physics*,[101] where he says, as dealing with a problem: 'Perhaps place is not a substance in itself, but is predicated in relation to the order and position of bodies, according to their natures and powers, equally in the case of animals and plants and, generally, of things composed of different elements, whether animate or inanimate, that have a natural shape. For the order and position of these parts is relative to the whole being. Therefore each is said to be in its own space through having its proper order, since each of the parts of a body would desire and demand its own space and position.'

[100] Aristotle denies that perpetual existence counts as being *in* time. For analysis, see R. Sorabji, *Time, Creation and the Continuum*, London and Ithaca N.Y. 1983, 210-17.

[101] = Theophrastus fr. 149 FHS&G. I am grateful to R.W. Sharples of Project Theophrastus for the full reference.

I will show that the divine Iamblichus bears witness to the same position. For he also, in the fifth book of his commentary on the *Timaeus*, chapter 2, wrote as follows:[102] 'Every body, as being a body, is in a place. So place came into being in union with bodies, in no way cut off from their first emergence among beings, nor from being in its central sense. So the *Timaeus* reasonably introduces place first of all, together with the coming into being of bodies. Those who do not treat place as being akin to a cause but reduce it to limits of surfaces or empty spaces[103] or even intervals of all kinds drag in alien opinions and, at the same time, miss the whole purpose of the *Timaeus*, which always joins nature with its fashioning.[104] As it introduced bodies at the beginning as akin to cause, so it was necessary to see place as connected with cause, as the *Timaeus* shows.[105] Also, just as we attempted to exhibit time as of like nature with fashioning, we should explain place in the same way.'

So Iamblichus himself rejected other approaches that make place something unconnected with things in place, and says that place is of like nature with things in place. Continuing, he says: 'What view, then, clearly states the perfection of place and its kinship with being? It is that which treats place as a power of corporeal character which sustains bodies and holds them apart, raising up those that have fallen and uniting those that are scattered, filling them up and surrounding them on every side.' So he seems himself to give a definition of place in which he calls it a 'power of corporeal character', akin to what is in place, 'which sustains bodies and

102 = Iamblichus, *Commentary on the Timaeus* Book 5, fr. 90 (Dillon); comm. possibly on Plato *Tim*. 52AB. An important definition of space as corporeal *power*. As J.M. Dillon points out, *Iamblichi Chalcidensis Fragmenta*, 60-2, the past tense used in 639,35 'as we attempted to exhibit time' confirms that the traditional order of Platonic commentaries was first on time, and then on space (with additional evidence from Calcidius).

103 Those who reduced place to surfaces are the Aristotelians. The one unclear item here is Iamblichus' (ap. Simplicium) accusation that some unspecified thinker(s) reduced place to *diakena khôrêmata*, empty spaces. Usually *diakena* refers to interstitial voids (Plato *Tim*. 60E). However J.M. Dillon, *Iamblichi ... Fragmenta*, 385, finds no parallel for *diakena khôrêmata*. Probably Iamblichus is attacking Epicureans, or the Atomists more generally.

104 Fashioning: the making of the world by the divine craftsman. The usual translation is 'creation', but the word is better reserved for the biblical creation *ex nihilo*. (J.O.U.)

105 Place connected with the Cause, Plato *Tim*. 30A, 36D, 52BC.

holds them apart, raising up those that have fallen and uniting those that are scattered, filling them up and surrounding them on every side'. It is also clear that, if place fills bodies up, it cannot be separated from what is in place. It surrounds bodies in such a way as to bound them and unify them.

Why, then, should we say that so many great men were mistaken in their opinions about place, putting forward our problems as an unfortunate feast for those who are accustomed to take pride in the apparent contradictions of the men of old? Should we not rather follow up each of those who wrote about place and show that none of them missed the truth about place? But, since place has many aspects, we should show that each man has seen and revealed a different aspect of it.

But now, in order that we may both see the common conception of the whole of place and observe from that standpoint the above-mentioned differences between the philosophers, I shall continue with a fresh start as follows.[106] That which is truly One as being in every respect an existence unextended and without inner distinction, and timeless, is the standard and measure of all things, needing no standard or measure for itself from anything. But the unified one is also controlled by the One and meanwhile has no need of such a completion. I am now calling unified not the continuous, for that is already extended, but that which is not subject to any separation into parts. What is separated has already suffered this in many ways. Some things became extended in virtue of their being or potentiality, like the so-called continuous quantities, some are divided by their plurality, like what are called discrete units. Some things were stretched out in accordance with the extension of their being, namely those that we say are perpetual or have their being for some period of time. Those things that totally departed from the limit of unity and moved towards the unlimited and unbounded were in need of measures and boundaries to preserve them as bounded and unmuddled. Of these, some tended to pour out without limit, others were in danger of their already divided

[106] Simplicius' account follows the typical Neoplatonic exegesis of proceeding from the One to the Many.

parts[107] being muddled together. In the case of the discrete plurality, a number containing the distinction within determinate forms stopped an unlimited scattering. In the case of extended quantity a measure of size, of which the yard or inch is said to be an image, checked its boundless spread. In the case of extension of existence, the unchanging magnitude is measured and united into one by eternity, the changing is united into plurality by time. These are the sorts of measure that did not permit quantity to be drawn out into unlimited indeterminacy, without distinction either in the paradigms or in their images.

There is another agent that stops the muddling of the distinct in all these cases, so that their parts, through their own completeness, are not muddled together, but each receives its appropriate order and position. For it is not without reason that among numbers the unit is set before the dyad, the dyad before the triad and so on.[108] Nor is it by chance that odd and even numbers are ordered alternately and consecutive square numbers have the intervals between them increasing by two,[109] and the other elegant theorems that have been handed down about the essential ordering of numbers in arithmetic. In the same way the distinction of the other incorporeal forms is determined. The same agent determined the disposition of the parts of animals and plants, in accordance with which the head, hands and feet have such and such a relative position and the roots, trunks and branches of plants this or that. It did not only determine the order of the parts in relation to each other, but it also settled the position of each whole thing in the wider whole. Thus the earth was allotted the middle of the whole, the heavens the circumference and intermediate things intermediate positions.

[107] Reading *merôn* instead of *monôn* at 640,34 as Diels in apparatus. Cf. 641,9. (J.O.U.)

[108] (Neo)pythagorean mathematical metaphysics, which can be found in Theon of Smyrna (second century AD) *de Utilitate Mathematicae*; Nicomachus *Theologoumena Arithmeticae*, in Iamblichus *de Communi Mathematica Scientia, in Nicomachi Arithmeticam Introductionem*; and even in Damascius *Principiis*. See D. O'Meara, *Pythagoras Revived: mathematics and philosophy in late antiquity*, Oxford 1989.

[109] Thus,

| | | | | | | | | | |
|---|---|---|---|---|---|---|---|---|---|
| of the integers | 1, | | 2, | | 3, | | 4, | | 5 |
| the squares are | 1, | | 4, | | 9, | | 15, | | 25 |
| whose intervals are | | 3 | | 5 | | 7 | | 9 | |
| whose increases are | | | 2 | | 2 | | 2 | | |

(J.O.U.)

Perhaps this has already been said. But now it should be added to the above that there is a common conception of the whole of place which says that it is the determination of the position of each distinct thing among entities. The distinction is in relation either to the receptacle or to the container or to the ordering of the position of each in relation to the rest. All these are observable in both incorporeal and corporeal things. For the intelligible orderings were allotted to the different receptacles of the intelligible universe as different places. At any rate Orpheus says about him who regulates the differences of dwelling-places:[110]

> Seizing such a measure, he divided out the universe to gods and men.

And THERE the container is often called place. That is why the Syrian Atargate is called the place of the gods, and Isis similarly called by the Egyptians, since they include in themselves the characteristics of many gods.

Plato calls that place in which the order of the intelligible forms is determined and distinguished in accordance with the order of the position of each towards the rest supramundane and intelligible.[111] With the same signification we say that each number is allotted its proper place when the unit is first, the dyad second, and next the triad, and, as was previously said, the square numbers have their areas and others have others in due order. Also the soul is called the place of forms in accordance with this form of containment.[112] Again, the special types of argument in dialectical proofs are called topics,[113] such as that from opposites and those from similars and those from genera and species.

Similarly, in regard to bodies, some people have thought that matter was place, being determined either by a

[110] Orphic Hymn, which appears also in Syrianus *in Metaph.* 182,10-16; for source cf. *Orphei Hymni* (Quandt) 79.t,3-9.

[111] Plato ref. to place of Forms, supramundane and intelligible: *Rep.* 508B13-C1; 509D1-4.

[112] Plato calls soul place of (intelligible) Forms: *Rep.* 508C (soul as the eye, and the light of the intelligible place), 517B4-7, 526D9-E4.

[113] Aristotle's *Topics* (*topoi*, i.e. places). The types of argument were memorised by being assigned *places* in the mind's eye: hence the name.

receptacle or by a container, since the one is naturally joined to it and the other outside it. In accordance with the theory of natural unity they call the shape of the surface the place, since the whole is within it; in accordance with the theory of external containment some have held the limit of the container to be place, since those things that have a position are inside it, as Aristotle did; others have held that place was an incorporeal interval, in the expectation of avoiding many anomalies alleged against previous views.

But the appropriately stated definition of place seems to me to be that it is incorporeal but perceptible and in bodies, being the ordering of the position of each thing in relation to others. This, I said, was the view of Damascius, the head of our school, and was vouched for by Theophrastus and the divine Iamblichus. On this view, just as a position is incorporeal, since position is neither the body situated there nor what it is situated in, so the determination of the position is incorporeal. Therefore differences of place, I mean above and below, to right and to left, and in front and behind, are incorporeal, even if they be seen as existing in a body. For we call a hand the right hand, not only on account of its different contour but also because it is situated on the right side of the body, like the liver.

Among incorporeal things, let us call the place of numbers their order that determines their relative position, and similarly in the case of forms, both intellective and psychic. But, indeed, in the case of bodies, as said above, place is that which provides for each part its order and proper well-positioning in relation to the others and to the whole, and also that which within the common whole, as within the whole universe, provides the appropriate position within the related whole. It is place in the latter sense on which Aristotle set his gaze; in this sense, one body is situated within another, a whole within a whole, as I am in the air, and the partial within the whole, in spite of the fact that an immediate place should be equal to that which is in that place. For he says: 'We 643,1
say that a thing is in the heavens as in a place, because it is in the air and the air is in the heavens. We say that it is in the air not as being in all of it but because its extreme and its container is in the air' [*Physics* 211a24]. A place of this sort

does contain and receive, in so far as it is greater.

Place in general is so in virtue of its demarcation of position. Thus Plato called matter the place of the forms and their space,[114] as receiving the position of the forms that descended into it and determining their extension. For the forms as such maintain the distinction of their parts in unity, but, when extended through the indeterminacy of matter, they assumed a material order of extension, just as previously they were distinguished by an immaterial order. So matter is the place of the forms both as a receptacle and as the cause of their extended position and extended order. But intervals are also place (*topos*), as being a preliminary outline and mould (*tupos*)[115] of the bodies which come into them, one type individual, as that of the heavenly spheres, one common, like that of the elements that mingle with each other. For in the same interval there may be fire, air, water and earth, one after the other, through the suitability of the interval for all of them. Nor is there anything marvellous if a less material body be the place of a more material body and be superior in its nature to the weaker, as the philosopher Proclus held. For the famous philosophers agree that the heavenly spheres are whole and full right to the centre.[116] It is also clear that the interior spheres are in the exterior, as that of the moon is in that of Mercury and that in the sphere of Venus and so on. Moreover, sublunary elements, animals and plants are in all of them, even if they are more immediately in that of the moon, since they are more akin to it. So each of the more widely inclusive is a place to that which it contains as a container and receptacle for it and as determining its position.

The Assyrian theology has handed down that there is another body, the aetherial, more divine than and beyond the universe, as Orpheus also knew, saying:[117]

[114] Plato called matter the place of forms and their space, room: *topos* and *khôra*, *Tim.* 52AB.

[115] Place (*topos*) = outline and mould (*tupos*), see Simplicius 611-18 on Proclus, 645 on Damascius.

[116] This is the problem of heavenly continuity and contiguity, and the doctrine of nested spheres; see E.J. Aiton, 'Celestial spheres and circles', *History of Science* 19, 1981, 75-114.

[117] Orphic Hymn on ineffable aether above heaven: Proclus *in Tim.* 1.176,13; 3.208,30; identified with the intelligibles and the Meadow of the Platonic Myth (*Rep.* 614-15), Proclus *in Remp.* 2.163,24. Cf. *hupsiphanês Aithêr, kosmou stoikheion ariston*, *Orphei Hymni* (Quandt) 5.t,1-4.

> Surround all things with ineffable aether, and place the heavens in the middle.

The astronomers also know this, since they necessarily place the starless and truly fixed sphere above that which is called fixed. For they believe that this sphere which contains the multitude of stars moves from the west by one degree in a hundred years, and thus they need a sphere to carry round the visible revolution from the east.

Since, as I began to say, Proclus had much confirmation of there being things superior to and with a more divine body than this universe, he reasonably took that body to be the place of the whole universe. It is clear that each of such bodies that receive and surround the others will be extended and have differences of parts and its own place, and thus will both be an interval and will also set the order of parts. As I said at the beginning and will now say again, each of those who discussed place set his eyes on some true conception of place and did not miss its general character. But it is not remarkable that all did not hit on all the differences of place. It is quite probable that there are other differences that have still not been brought to light.

Let me add this to what has been said. Our colleague Damascius first grasped well the concept of place as good ordering, and he analysed it first of any known to us. But he did not pretend to have discovered the other concepts of place, either those regarding the incorporeal or those regarding the corporeal. For I think that it is well said that place is the determination and measure of position, but even among incorporeal things there will be position according to their order, as, for instance, we say that numerically the dyad is situated before the triad and the triad before the fourfold. There is also a position among bodies, according to the difference of their extension.

I also draw attention to the fact that when Damascius posited three measures, number of separate being, time of endurance of change and place of extension, he did not distinguish the two senses of 'extension', one of determinate magnitude, according to which we say that one thing is a yard in length, another an inch, the other of extended position, according to which we call one thing above and another below, one to the right and one to the left. Though he often said that place was the measure of position, he never distinguished it from the measure of magnitude.

It would not be a bad thing to listen to his words in his book *On Place*, where he writes:[118] 'It has become clear to us that there are three measures of three types. There is no other type in the world of change except these three, and no other measure. For the indivisible is also threefold; there is the monad, the now and the point. The material division of the point, or, to speak more accurately, the individual, creates plurality, which is delimited by a certain number, since the single thing or monad is what is undivided in number. The flow of the now creates endurance in time. I call a now that which, in the flow of that which flows, is itself without flow, such as the impulse to change that is the beginning of change. It was allotted to time to measure the temporal now and the flow, as plurality fell to number. The expansion of the point produced extension, with which place coexists as the measure that determines the total position of the whole, in so far as, being extended in three directions, that is, in all directions, it will ensure the well-situating of the whole in regard both to its own internal omni-directional position and to all the parts in it, and, further, in regard to the position of all the parts, both in the totality of place and internally. Consequently, if something were a sphere or had a different middle and boundary, it would be situated in place where it would be well-situated.

It is clear that he says that place is the measure of position, well-disposing what is situated there. But he defines it not only as the measure of this but also of magnitude *qua* magnitude. For he says shortly afterwards: 'It is like a preliminary outline of the whole of position and of its parts and, as one might say, a mould into which that which is situated there has to fit, if it is to be properly situated and not be muddled up in an unnatural state. Examples would be a whole larger or smaller than the proper size, or situated in the

[118] Damascius on three definitions of place: absolute (*kuriôs*) and superior to contents, relative (*pros ti*) and equal to contents, and relative inferior to contents, *Principiis* 2.219,18-220,1.

whole where it is not seemly, or whose parts do not have their proper position, like a man with his brain in his heels, as they say.' Thus clearly he judges place to be concerned not only with well-positioning but also with the size of an extended thing being greater or less. But I think that he should have defined four measures, number, magnitude, place and time, the first measuring division, the second extension, the third all kinds of position, the fourth endurance in becoming. So I beg the pardon of scholars for these reflections on place and return to what comes next in Aristotle; I shall be concerned with the void.[119]

[119] At this point Simplicius returns to the detailed examination of the text of Aristotle's *Physics* at 213a12.

# Corollary on Time

## Contents

# Simplicius: Corollary on Time

That must suffice for the clarification of Aristotle's discussion 773,8
of time. For I have gone through what he has said from the beginning to the end, showing how it all fits together so far as possible. But since our academic training has not the sole aim of learning Aristotle's opinion concerning the nature of time, but rather of understanding what time is – by which I think that we shall also get a closer grasp of Aristotle's thoughts on time – let us examine that briefly. After that, let us review the opinions of those who have philosophised about time. But since Aristotle at the beginning of his discussion mounted severe attacks on the reality of time and did not defeat these attacks it would be useful towards our goal to overcome them so far as possible. For otherwise our discussion of time would remain incomplete.

Now the unitary and unified nature, remaining in its simplicity, rises above[1] all differentiation. The unitary nature is one, the unified is subordinate to the One, having diverged from the One a little towards being a single existent. Now there is THERE no differentiation of essence from existence.[2] But

[1] *huperanekhei* is a Neoplatonic term, Proclus *in Alc.* 139,7; 162,9; *in Tim.* 2.5,7; 3.201,24: it means to rise above, to a level beyond. For Neoplatonists such transcendence is a measure of the unity, wholeness and reality of that entity or property. Below the One each level increasingly displays differentiation, plurality and extension. The relations 'above' and 'below' are not only metaphysical; for the Neoplatonists they denote logical priority: the simple comes logically before the complex. The increasing differentiation is the result of the 'unfolding' of wholes into parts (see n. 86 to the Corollary on Place). So while the whole itself 'remains' unchanged, its parts 'proceed' down (in the converse direction, the 'return' from part to whole is a reduction, an ascent from complexity to simplicity).

[2] 'THERE' (i.e. in the intelligibles) essence is not differentiated from real-existence (but at subsequent levels they are differentiated, 773,26f.). S. uses the infinitive *to einai* (to be) to denote essence, and the participle *to on* (being, that which is) to denote real-existence. The meaning of this abstruse passage is that at the intelligible level unified wholes are in essence 'really real', or real-existent (see n. 1 above). This is because the essential nature of intelligibles is to be changeless, permanent, defined

where within what is unified such a differentiation has appeared, there plurality has appeared together with unity, and a whole with parts has come into being, and some preliminary outline[3] of things HERE that are spatially separated has made its appearance, and essence has become something other than existence. But where any sort of differentiation has appeared in things *qua* unified, there had at once to be some measure of the difference, in order that what departed from unity might not fall into indeterminacy. Therefore the manifold received its measurement by number, in order that it might not be undenumerable and truly unlimited and indeterminate. Also the continuously divisible was measured by the criterion of size, by which it was of such a size, and thus the continuous, like the discrete, resisted the fall into indeterminacy. Likewise the spatial distinction of parts one from another was bounded by proper arrangement and ordering in place. So these are the measurements where there is differentiation between essence and existence.

But since the essence of the existent, parallel to the distinction among the unified, has become like an extended life of the existent, but has none the less remained in that which is one, because actuality lies in essence (for the essence of things THERE is actuality, as Aristotle himself saw with inspired insight),[4] it accordingly received as its measure

---

and fixed by mind: the features of (Platonic) reality. However, 'HERE' on the changeable, physical level things cannot be fixed by the mind and are indeterminate. They are appearances, phenomena, and their essence is no longer real-existent.

[3] The 'unified' is defined as plurality mixed with unity, and distinguished from 'unity' itself. According to the Neoplatonic account of the procession from unity to plurality, the One itself is beyond any kind of definition or description, and beyond being itself. Below this the first real existent, being, is a mixture of unity and plurality. The single 'existent' is more complex than, and distinguished from, single 'unity'. The first Being may have a 'unified nature', but cannot be simply one: it contains potentially the manifold of beings that make up the universe. Thus it is fittingly the 'preliminary outline' (*proüpographê*, 773,25; cf. Plotinus, *Enneads*, ed. Armstrong, 6.7.7), viz. the Platonic Paradigm of the universe (*Timaeus* 29BC). Proclus discusses the first, intelligible, Being in the *Platonic Theology* Book 3. Pr. is also Damascius' source; Simplicius follows Damascius, or draws directly on Proclus.

[4] In the intelligibles essence is always actualised, cf. Aristotle *Metaph.* 1088b26. That Simplicius uses both Plato and Aristotle is not at all strange. Since Porphyry (third century AD) the Neoplatonists sought to unify the diversity of Greek thought, and particularly its two leading, but antagonistic, authorities, Plato and Aristotle. This 'harmonisation' was not a symptom of feeble thinking. It was the critical evaluation of existing philosophies. In analytical fashion, only the ones that were

eternity, which brings the extension of being into unchanging rest in that which is one. Thus appeared those four measurements of the fourfold differentiation of the intelligible – number, size, place and eternity.[5]

But the perceptible proceeded from the intelligible, and the generated from that which is, by some side turn and divergence;[6] for it did not remain within imposed measures of being, but was contaminated with not-being, accepting the corruption away from being in not-being for the sake of some sort of existence. And that is why becoming was given the substance of an image of being. So the perceptible, as one might expect, did not, as THERE, admit a formal distinction, by which each thing was not only unified with all other things but had a real being like them. The becoming admitted rather a distinction imposed upon it, emerging as a composite thing from the incomposite nature. Because of this, what THERE is a distinction of the manifold is HERE a rending apart, and the natural wholeness THERE, breaking HERE continually apart, appears as enmattered quantity.[7] And what is THERE a distinction of parts has HERE become a scattering asunder. Thus also the extension of being, which is THERE distinguished formally from being but which remains within the single being in reality, is HERE extended and derives its nature from its participation in the creative process.

---

deemed coherent were accepted, the rest were rejected or modified radically. Aristotle's logic and some of his theory of change and metaphysics was incorporated into Neoplatonism, but his science and particularly the doctrines of the aether and the celestial bodies, of place, motion of the four elements, were rejected in favour of theories based on Plato (and the Stoics). Following Iamblichus (fourth century AD), the Neoplatonists taught both Aristotle and Plato, in that order, as from a particular to universal. L.G. Westerink, *Anonymous Prolegomena to Platonic Philosophy*, Amsterdam 1962 (revised French edition, in Budé series, 1990).

[5] Four measures, see the Corollary on Place, 645,15.

[6] 'To proceed' (*proelthon*),'side turn' (*ektropên*), 'divergence' (*parallaxin*): all these terms refer to the Neoplatonic scheme of procession from the One to the Many, and from real being to 'iconic' being. The real being is the Paradigm, that is, the Exemplar or Standard. The entities down below in the series are not substantively identical to it, but only *like* it. They are copies of the original, and images ('icons') of the really real (for the influence on Byzantine theology and iconoclasm, read Maximus the Confessor).

[7] Enmattered quantity (*enulos posotês*), cf. Proclus *in Parm.* (Cousin) 1123,7, as the eighth level, pure quantity abstracted from body co-terminous with the universe. This would explain its juxtaposition with 'whole-being', which is the universe as a whole.

Furthermore, the measures of imposed divisions themselves suffered a descent into becoming, number being torn into its units, the measure of size divided, place being shared out into portions and time flowing and separated into before and after. And so these measures participated in becoming so that they themselves seemed to need further measures. For number HERE seems to be numbered and the measure of size, such as the yard, to be measured, and the portions of space seem themselves to need arrangement and ordering and positioning, and time seems itself to be in time, as we shall see as we go on.

At this point I beg to differ from my tutor Damascius, who would have it that there are not four modes of measurement, but three – number, space and time – in his work entitled *On Number, Space and Time*. 'As number is the measure of plurality', he says, 'so place is the measure of size.' But I think that it is perfectly clear that there is one mode of measurement when we say that something is two or three yards long, and another when we specify an ordering according to which one part is above, one below, one to the right, one to the left.

So time is the measure of the flow of being, and by being I mean not only essential nature but also activity. Aristotle saw wonderfully well the nature of time and made it clear, saying that for process 'and the rest to be in time is that their being is measured by time' [*Physics* 221a8]. But just as a process does not take place in indivisible parts (for it is not composed of elementary changes, nor a line of points, but the limits of both a line and a process are indivisible, whereas the portions of them of which they are composed, being continuous, are not indivisible but divisible), so in the same way some elements of time that are bounds, the 'nows', are indivisible, but the portions of time are not so. For, since time is continuous, it too has portions that are infinitely divisible. So that, even if process and time be in continuous flux, they are not unreal, but have their being in becoming. But becoming is not simply not-being, but is to exist at different times in different areas of being. For just as eternity is the cause of that which undergoes an intelligible differentiation from its own proper unified being remaining within its own single being, so time is the cause of the dance around[8] the intelligible One[9] by that radiance from the form which has descended from THERE into perception and which holds the continuous dance in order. For just as because of place the portions of separate things do not merge together, so because of time the being of the Trojan war is not confounded with that of the Peloponnesian war, nor in each person the being of the baby with that of the adolescent.[10] It is clear that everywhere time is involved with process and alteration, holding together in becoming those things which have their being therein, which is the same thing as to make that which becomes dance around that which is.

Damascius has well said that time, both everlasting and occasional, controls in their becoming and holds together all generated things and all generation. He goes on to say that 'in itself time would be a cause of changelessness in things which of their own accord depart from being what they are, so that it is rather a cause of rest than of change'. This seems to be said because of the analogy between time and eternity and the continuity in the world of becoming (for as the latter is the cause of permanence in being, so time is the cause of permanence in becoming). But complete permanence is never suited to time, just as being is not to becoming, but just as becoming is a development departing from being, so the dance around being is a development from remaining in being.

But those things that he says do not disturb me so much, but rather those things that he often said to me when he was alive,[11] without convincing me, to the effect that the whole of

[8] Time described as a dance (*khoreuein*), see also Simplicius *in Phys.* 705,5-9; as the dance of the now 786,30-3. In the broad Neoplatonic context the 'dance' is the cyclic process of procession and return around that which remains. More specifically, time as a circulation derives from the celestial bodies as cosmic clocks, Plato *Tim.* 77ff. (time created with the heavens).

[9] The intelligible one is distinguished from *the* One. The 'intelligible one' is a qualified one, since Being is a mixture of the Limit and Unlimited in Plato's *Philebus*; cf. Proclus *Pl. Th.* (Saffrey and Westerink) 3.85-7.

[10] Time distinguishing events, cf. Corollary on Place 626,14; also Simplicius *in Cat.* 356,29-30; 364,15-17; *in Phys.* 1160,28-30.

[11] This is most significant for the dating of *in Phys.* and for the end of the Athenian Neoplatonist scholarchs. First, it tells us that this was written sometime after Damascius' death. I. Hadot, *Le problème du néoplatonisme alexandrin: Hiérocles et Simplicius*, Paris 1978, 20-32, calculates it to *c.* AD 537/8. Second, this reference tells us that Simplicius was evidently in a position to have heard news of Damascius' death. At minimum it means that Simplicius lived in a place where he could be

time existed simultaneously in reality. I wish to expound and examine this thesis, which is also in his work *On Time*.[12]

Aristotle said that a thing was in time if it was possible to find time external to it (for thus it was encompassed by time).[13] Objecting to this Damascius wrote as follows: 'One might be still more surprised on hearing the definition of a thing in time as that which has time external to it (for thus it is enclosed by time as its container), which time measures and counts, being outside the limits of what it contains, like place. For this is what Aristotle says. But, if this is so, how can time, as a measure, be an affection of process? For place has not that character. And if time is a measure, the temporal object has been measured in time even if one can find none outside it. There is the further absurd consequence that the totality of time is not time. I do not mean the form, or the shape, which Aristotle himself said was all simultaneously time, just as one might say that Socrates was the whole of man at once, but the common ever-flowing time, which he says is contained within eternity. Yet we agree that the ever-flowing time is the same as that in which the everlasting alteration of the forms occurs and the everlasting motion of the heavenly bodies. But according to the definition these are not in time. For what could one conceive of external to ever-evolving time? It is also paradoxical that the parts of the whole of time do not constitute time. For, if the past, the present and the future are parts of time, or if last year, this year and next year are parts of time, then the whole composed of them ought to be a time, whose name is everlasting. It is also strange that that which is ever numerically identical should not be in time because its being is not in flux and is naturally outside flux. Aristotle says that only such things as the incommensurability of the

---

contacted by Damascius' messengers, who knew his address. If so, this was perhaps not the only occasion in which D. and S. communicated. If they were back in Athens, or in their respective home provinces in Asia Minor, regular communication between them would have been easy: their freedom had been guaranteed by the Eastern Emperor as part of the deal with the Persian King Chosroes. If, as M. Tardieu claims, Simplicius settled in the no-man's-land of Harran, then communication would have been extremely difficult (though not impossible). S., however, does not tell us explicitly whether he was in regular contact with his old teacher and colleague.

[12] On Damascius' view that the whole of time exists simultaneously in reality, see further 777,1; 798,4 and Introduction.

[13] Aristotle on things measured by encompassing time: *Phys.* 251b11-14; *Metaph.* 1021b12-16; no time external to universe, *Cael.* 279a10-20; 304b28-31.

diagonal and the side are everlasting. But if the longest lasting – the multitemporal – is in time, why not also the totitemporal, to coin a term? And why do we say that what is everywhere, such as the universe or, if you prefer it, the sub-celestial universe is in place and indeed the whole of place (for where is the whole that is everywhere?), but in the case of time do not recognise as being in the whole of time that which is everlastingly numerically identical? For as the everywhere is to place, so is the everlasting to time. And why is place perpetual, such as the hollow periphery of the heavens and the centre of the universe, and number which is ever the same such as that of the stars, but there is only one time, that comes to be and passes away, beyond which there is always more? For if there is always a further number, none the less there is no perpetual number outside perpetual number, and every time is a number.'

On this matter Aristotle maintains a similarity of time to place, in so far as time bounds and contains what is in time, 777,1
just as place does what is in place and number what is numerically described. But Damascius claims that, if one cannot find place outside the whole of place, so one cannot find a time outside the whole of time. But yet it is clear that Aristotle also believes that place is different from time in that nothing prevents place from being at any time complete, whereas time has its being in becoming, like process also, and so does not exist complete at once. If, then, time has its being in becoming, as does process, what is there absurd in the whole of time not existing at once? The opposite would have been absurd – that the whole of what has its being in becoming should exist at once. If then it is not absurd that the whole of process does not exist at once, because it is a flow, so it is not absurd that the whole of flowing time does not exist at once, even if the common form of endless repetition remains, as in the case of process. If we speak of the everlasting flow of time, we mean by 'everlasting' not a simultaneous unlimited existence but unlimited continuation. Thus we say that both time and process are everlasting, as never being exhausted, not as a simultaneous whole lasting for ever. For the everlasting is of two kinds, in one way meaning a simultaneous whole, like the eternal, in the other as having

its being in becoming throughout inexhaustible time. In this latter way, therefore, also the everlasting alteration of the forms takes place, which itself flows in its being, but remains in form through its continual recurrence, as also the everlasting process of the heavenly bodies. Just as the motion of the heavenly bodies is not a simultaneous whole, nor is the sum of time which measures it. And on this account we cannot find anything outside the simultaneous whole of time, because it does not exist, but of the time that is ever in existence one can find time outside because of its inexhaustible flow. But also I do not think it paradoxical that the parts of time, which do not exist as a whole simultaneously, should constitute times like the past, the present and the future. For if we said that the parts existed simultaneously, it would be strange that the whole did not exist simultaneously while the parts did. But if they are parts in this way, as of that which is not yet and that which is no longer, how could the whole exist simultaneously? Perhaps one could write more clearly to exhibit[14] the paradox: if the whole of time exists simultaneously, it is paradoxical how the parts should not exist simultaneously; and if the parts do not exist simultaneously, it is a paradox if the whole is simultaneous. But the term 'everlasting' applies to time so as not to imply that the everlasting is simultaneous, as I said before, but as being everlasting in the sense of inexhaustible.

Damascius says that the perpetual things are implausibly stated by Aristotle to be atemporal in the same way as the incommensurability of the side and the diagonal. Yet, if this is a case of something actual and complete at one time, it is clear that it is a case of being and not of becoming, just like the dyad being even and the forms incorporeal, so that these would not 778,1 be in time but rather in eternity. But it is clear that whatever is in time is in process and will not exist complete at one time. For we shall not cause supratemporal things to be in time when we make acquaintance with them in time.[15] 'But', he

[14] Reading *houtô deixoi* instead of *egô deixô* at 777,29 as Diels in apparatus. (J.O.U.)

[15] The Neoplatonists identify metaphysical superiority with logical and cosmological priority. In other words, as the fundamental principles come before the secondary, and the premises before the conclusion, so the more complete and real existence comes before the less complete and phenomenal. Therefore the

says, 'if what lasts through much time is in time, why not also what is in all time?' Let us agree that this too is in time, if it is in process. Then it cannot exist as a simultaneous whole. The reason why we say that the circular motion is perpetual is that it itself is inexhaustible and is measured by inexhaustible time. Also there is no necessity that if there is something simultaneously in the whole of place, which we call everywhere, like the whole of the bodily universe, that there should also be simultaneously in the whole of time something which is everlastingly numerically identical. For there is nothing odd in the whole of place existing at one time; for place does not have its being in becoming as does process. But how could this be so in the case of time which is the measure of process? So even if there be an analogy between the everywhere and the everlasting, still the former remains constant, the latter flows. Also, if everywhere happened to be boundless, it would be an unchanging boundless, but time would continue boundlessly. 'But why', Damascius says, 'are place and number perpetual and prior to the perishable, but not time?' Perhaps time also is perpetual (for Aristotle shows that time is in any case inexhaustible); but it will be so having its being in becoming, just like process, and they will both be everlasting, but not as being a simultaneous whole, but as proceeding without limit.

But let me draw the threads together and give my own opinion. If there is some coming to be and process which does not have its being in becoming nor each of its parts one after the other, but exists as a simultaneous whole (as one might say that the being of the psychic, or, if not this, then at any rate the being of the celestial body was generated as existing from a single cause and thus in change in virtue of the process of its coming to be, through which it departed from true being into becoming and from being a pattern to being a copy),[16] if then someone looks towards this sort of genesis and process, as my friend Damascius often maintained, he has also a plausible claim to envisage also a time which is a simultaneous whole and measures that sort of process and the being of what exists all at one time. But if Plato defines the

---

supra-temporal comes *before* the temporal.

[16] Pattern (*paradeigmatos*) and copy, icon (*eikona*), Plato *Tim*. 29B, 31A, 48E-49A.

created as what comes into being and passes away and never truly is,[17] and if every process is alteration and always contains one thing after another, it is not right to call created or in process things which are simultaneous wholes, unless it is possible to see in some way alteration in them. Now I know that Damascius, in extending coming to be and passing away to include perpetual things which were created but unchanging in their being, said that their nature was between being and not-being, and that he said that they were coming to be *qua* striving towards being and passing away *qua* descending towards not-being; but perhaps the intermediate, such as the nature of the soul, is not truly created but is between the indivisible and the divisible, that is to say, between being and becoming, and on this account is measured by some measure intermediate between time and eternity which, since it has no name of its own, the philosophers sometimes call eternity and sometimes time.[18] But the being of the heavenly bodies is wholly created, for it is visible and tangible and bodily.[19] Being created, and not having remained within the one, it became subject to process and alteration, and, even if not to change of nature so as sometimes to exist and sometimes not, then some type of change through which the heavenly bodies in their different configurations take a share of a different life and a different perfection into their

[17] The 'generated' (*genêton*) in this sense: that which belongs to the temporal realm of generation and perishing (coming-into-being/becoming and passing-out-of-being), where things are perceptible by the limited, physical senses, Plato, *Tim.* 52A; but cf. 779,1, soul 'not truly generated' (*oude genêton esti kuriôs*), whereas, 779,5, the celestial body is 'generated' (*genêtê*). The difference between 'creation' (*dêmiourgia*) and 'generation' (*genesis*) is that the former encompasses the latter. The Neoplatonic Creator, Intellect, produces Soul, Nature and Body (intellective forms are also products of the Intellect). Soul and Nature (as hypostasis) are immaterial, neither changeable nor perceived by the senses. Therefore while all souls, natures and bodies are 'created', only the entities with material bodies are strictly 'generated'. The four elements occupy an intermediary place, because as parts they are mutable but as wholes immutable.

[18] Measure intermediary between eternity and time (proper): Proclus *El. Th.* proposition 191, had pointed out that soul has an eternal substance, but temporal activity. Soul stands between the purely eternal intelligibles and the purely temporal bodies. Proclus had also distinguished between Time Unparticipated, viz. Time which 'remains', and Time Participated, which is always in motion, viz. proceeding and returning. Further, he had distinguished between perpetual time, that proper to the unitary World Soul and the heavenly bodies, and the temporary, parts of time, *in Tim.* 3.19, 28; *El. Th.* propositions 51, 53-5 (L. Siorvanes, 'Proclus on the Elements and the celestial bodies', PhD diss. London 1987, 112-18).

[19] Heavens generated, visible and tangible, Plato *Tim.* 32B.

natures from their own paradigmatic forms and pass on a share to each other and to sublunary things.[20]

But since, when he makes a division of the sorts of time, Damascius says much about the time that exists as a simultaneous whole, we had better look into these views. This is the thesis: 'What is the whole of time? Surely, it is that outside which there is no time. But what example could we give of time outside time everlasting? To Aristotle the everlasting seemed to be supratemporal and akin to eternity; for he declares clearly in his discussion of time that one can find a time outside every time. Now the everlasting is the greatest time of all. But is that which is ever and ever becoming and time that flows the greatest time? But this everlasting will never arrive at being altogether nor become a whole all at once, and this everlasting is not everlasting, but rather never. For it is hard to imagine how the everlasting can be at some time, for in that case some time and the everlasting will be the same and simultaneous. But nor can the everlasting be always – for we say that this everlasting does not exist all at once, but flows one bit after another. Now it is paradoxical that some particular and minutest piece of time should exist, but never the whole of time. However, if the word 'everlasting' means something and is not a *vox nihili*, it is impossible for the everlasting not to be an everlasting object. The perpetual makes the point clear; for that is perpetual which cannot sometimes or always be non-existent. But what cannot be non-existent must necessarily exist. For if the everlasting does not exist, or not for ever, then nothing of things created will be perpetual; so that neither the heavens nor the universe will be for ever numerically identical, but everything will flow, since the whole of the everlasting will be in flux and will neither be nor become. For always some particular time will come to be and not the everlasting.'

For these reasons he claims that the everlastingness of the heavens and the whole universe is a simultaneous whole, not

[20] The heavenly bodies are changeable only insofar as they move in place; they are not perishable: hence a common title of them, the 'moving eternals'. Neoplatonically speaking, they contain the principles of change that prevail in the sublunaries. This is in addition to being the causes of change of seasons, weathers and climates, the physical astrology of Ptolemy *Tetrabiblos* 1.1, which is based on Aristotle *GC* 2.10, 336b and *Meteor*. 2.5, 362a-365b.

by proceeding without bound and continually coming to be in flowing time, but in a time which exists simultaneously as a whole. In what follows I think that he looks more accurately at the matter, having formerly elsewhere posited[21] the existence of such a time. He says that it derives from THERE and exists as such in the nature of the heavens. But it is necessary to add more. For he examines what it is that creates that which again and again comes into being and never ceases to recur and concludes: 'At least what is natural nature will create, and before nature soul; for soul initiates all change. And how can nature that is the maker of perpetual bodies and that pours from herself the ever-flowing time not be perpetual and full of perpetual principles?[22] So surely the perpetual priniple of time which has its being in nature should be this time which is all and as a whole ever present. Similarly the principle of time that pre-exists in the soul will be numerically the same time. But if in the unalterable soul and nature of this sort time is brought together into a whole, the form of time, then the whole of time will be for ever unchanging in everlastingness and in no way flowing; it will have in one the earlier and the later and that now which we call the present time. So from the stationary, indivisible time the time that flows will be divided into three in a certain way to be explained later.' Continuing, he says that one sort of time is in the activities of unalterable natures, activities that alter, whether in psychic or bodily things, and which preserve their everlastingness only in form. A second everlastingness is seen in unendingly changing natures, which is formally preserved in their totalities. A third type of time is involved in things that come to be and pass away, that is bounded and is either cyclic like the recurrence of the whole creation or perhaps the single circle of the sun from and to the same place, or else is rectilinear as is seen among simple and perishable bodies. That is the way he divided flowing time; but to the above he added the following: 'If, then, the creative agent is perpetual and in everlasting

[21] Reading *hupotithetai* instead of *apotithetai* at 779,36 with MS a. (J.O.U.)

[22] Nature pouring from herself the everflowing time: a Chaldaean Oracle's reference to Hecate (the intellective Rhea 'flowing' at a lower level), see Chaldaean Oracles fr. 54 (des Places), in Proclus *in Parm.* 821,7; *in Tim.* 1.11,21; 3.271,11. See also Porphyry *de Philosophia ex Oraculis* (Wolff), 174,1-4; *in Cat.* (*CAG* 7), 104,30-5.

time, it creates by its existence[23] the elements of the world of becoming. For if it creates by its activities, still more does it by its existence. For being is creative before activities, and the perpetual before that which is not such, so that through its being it implants also in those perpetual things among those that become the time which is everlasting and never flows, or rather will bind them to the perpetual with this time. For the being of the heaven, that participates in everlastingness became perpetual *qua* everlasting, that is, by being substantialised for the whole of time. So the time that has the same nature and source as the complete and perpetual, whether it originates from the soul or from its own nature * * *.[24] This is the time complete in its existence which neither flows to the later nor is associated with such a time, and from it the flowing time winds off part by part. Because of this time, flowing time never ceases to arrive and to end, since the real time that is ever numerically the same remains activating that which is everlasting in form. For as the activity of the heaven is an inexhaustibly recurring circular motion, so the time of the circular motion is the activity of the real unified 781,1
time. But how, someone may ask, does this time which we call unified differ from eternity? For Aristotle, at least, would call this time eternity, in which was enclosed each time as it came to pass. Very likely Alexander would follow Aristotle in saying this. Perhaps one might reply to Alexander that it should be considered whether Aristotle will treat as identical the being and everlastingness of the substance of the fifth body and that of the unchanging object of desire. For if the latter is the cause, the former causally dependent, as Aristotle agrees, and if he does not think fit to predicate generation of bodily perpetuity (for this is the very thing that the word "perpetuity" excludes), if, then, they differ, he would admit that everlastingness was of two kinds, one being cause, the other effect. So we who agree that what exists through something else is created and in becoming would reasonably not call this eternity, but time, counting it as the first image of eternity.'

[23] Creative agent creates by its very existence: this is the Neoplatonic concept of spontaneous creation due to the agent's own 'superabundance of potency' (cf. Proclus *El. Th.* proposition 121).

[24] Lacuna. (Ed.).

In this discussion, to begin at the end, if this time is something in becoming, it must be because it is in process. For where there is coming to be, there there is inevitably process and alteration. For Plato calls that which comes to be and passes away created and never having true being.[25] The first of all times, which he says came into being with the heavens, he calls a numerically moving image of unchanging unitary eternity,[26] conceiving process as necessarily involved in time. Damascius rightly asks Alexander and Aristotle before him to say how the being and everlastingness of the fifth nature differs from the being and everlastingness of what they regard as the first cause. I think they would reply that the latter kind of being and everlastingness was eternal as containing the boundless in unified form and such that it must be always taken[27] as a whole, whereas the being of the fifth essence was perpetual, not as a simultaneous whole but as proceeding without bound; therefore it was also in time. For whatever portion of its existence one may take will have something beyond it, since it advances without bound. If anyone were to call such a time eternity, wishing to call everything that is everlasting eternal, and dividing this into a simultaneous whole and that which has its being in becoming, then he would be in verbal disagreement with Plato, but would distinguish things rightly and in agreement with him.

But I consider it desirable to consider again what was said at the beginning of the passage just quoted, where he understands the everlasting in one way only, as the time that collects the boundless into a single whole. Now such an everlasting does not seem to Aristotle either to belong to eternity, as Damascius says. For Aristotle says in the third
782,1 book of the *Physics*: 'The boundless is manifest differently in time, in men and in the division of magnitudes. For in general the boundless consists in one thing being taken after another for ever, and what is taken is always bounded, but ever another and another. Again, being has many senses, so that

[25] See also 778,29. For Plato the generated is Becoming, *genesis*, viz. it never is true Being (*on*): *Tim*. 28A.

[26] Plato calls time the moving image of eternity, *Tim*. 37D, since time brings things into existence in conveyor-belt fashion.

[27] Reading *lambanein* instead of *lambanôn* at 781,24 as Diels in apparatus. (J.O.U.)

we must not conceive of what continues without bound as being like a particular, such as a man or a horse, but as we speak of a day or a contest existing, whose being does not come about like a substance, but ever in coming to be and passing away, bounded but for ever one after another. But in the case of spatial magnitudes this happens with what was taken persisting, but in the case of time and men they pass away without the supply being exhausted' [*Physics* 206a25ff.]. But he still more clearly explained this process without bound when he said: 'The boundless turns out to be the contrary of what people say it is. For it is not what has nothing outside it but that which always has something outside it that is the boundless' [*Physics* 206b33].

Damascius seems to attack as absurd this view that such an everlasting never exists as a whole and that the everlasting does not exist everlastingly. But what would be absurd in the case of such an everlasting that has its being in becoming? Also he says that another absurdity is that a particular time, however short, should exist, but the whole of time never. However, it would be more impossible and more absurd that things having their being in becoming should exist as simultaneous wholes. In general he says that, if the everlasting 'is not a *vox nihili*, it is impossible for the everlasting not to be an everlasting object. The perpetual makes the point clear; for the perpetual is that which cannot sometimes or always be non-existent. But what cannot be non-existent must necessarily exist' [cf. 779,25ff.]. This argument also seems to me sophistical, since it takes that which ever is instead of that which ever becomes. Nor is the everlasting *qua* ever becoming a mere empty word, but it is in existence just by ever becoming. For this too is signified as a kind of being, insofar as it participates in being in a way just by becoming. For the dance exists when it takes place and we say that the dancer dances, but not through the dance being a simultaneous whole, but because the dance has its being in becoming. Continuing, Damascius seems to wish that this everlasting which is stationary and numerically identical be also for created things, which is very surprising. 'For if the everlasting does not exist', he says, 'or not for ever, then nothing of things created will be perpetual; so that neither the heavens nor the universe will be forever numerically identical, but everything will flow, since the whole of the everlasting will be in flux and will neither be nor become. For always some particular time will come to be and not the everlasting.' Well, if he requires to view the everlastingness of the created heaven and the universe as static and as a simultaneous whole, let him say how that which truly is and is eternal will differ from the created.

783,1 But it seems that the invariable procession through an intermediate[28] is the justification of Damascius himself rightly asking how the heaven and the universe can each be numerically identical, if they have their being in coming to be and passing away, and of my replying again to ask how, if this generated heaven and bodily universe has a static everlastingness, existing as a simultaneous whole, how such created things will differ from that which is eternal and which has true being. For I think that two extremes are signified by the word 'everlasting': (1) the properly eternal that has true being and exists as a simultaneous whole; (2) that whose being is in flux and which contains at different times a numerically different process of becoming, as is clearly seen in the movement of the heavenly bodies and all the forms of things in the sublunary sphere.[29] For example, the form of man is continually seen in one and another individual who comes to be and passes away. Between these two extremes there are two intermediates, the psychic and that which is at once natural and bodily, the psychic adjoining true being, while the natural composition of the perpetual bodies adjoins that which is mere becoming. Of these the psychic, though it departed from the eternal grouping and the wholly indivisible being, still did not

[28] 'Procession through an intermediary' (*dia mesotêtos proodos*) is the characteristic Neoplatonic expression of the transmission of property through Mean terms from one to many, or from sameness to otherness. It stems, as I have noted earlier, from the Neoplatonists' identification of logical with metaphysical sequence. As conclusion 'proceeds' from premises through the mean term, which maintains the syllogistic continuity, so substantive entities follow each other without metaphysical 'gaps' (the absence of physical voids between bodies is no more than a case of this rule).

[29] The celestial bodies are the natural intermediaries between intelligibles and sensibles. According to J.M. Dillon, *The Middle Platonists*, London 1977, 49, post-Aristotelians focused on the division between the changing sublunary realm and the unchanging intelligible. This would explain why Aristotle's celestial aether becomes unimportant, because the celestial region occupies more of an intermediate status, and the division into sublunary-heavenly becomes secondary.

descend to the wholly divisible either in its nature or in the duration of its existence. That is why, while Plato and Aristotle both divided into two the measures of continuance of being, namely into eternity and time, Plato on the one hand placed both the intermediates in time, saying of the soul that 'it began a ceaseless and intelligent life through the whole of time' [*Timaeus* 36E] and of the whole heaven or the universe that time began with the heaven, whereas Aristotle rather placed both in eternity, attributing time to those things that visibly are always coming to be and passing away. Strictly speaking the measures would be intermediate between the intermediates and would appropriately have other names. Further the everlastingness of the psychic being and of the heavenly and cosmic wholeness must not be thought of as a simultaneous whole like the truly eternal, nor as coming to be and passing away like the condition of sublunary things. But in the same way as their essences have an intermediate nature, so does the continuance of their being and so do the measures of the continuance of their being. If anyone wishes to call these time, we need not quarrel about words; but let him not assign to them the sense of 'time' that is the common one divided into present, past and future to which both Aristotle and Plato assign the name. Nor should he recognise only that everlasting which exists as a simultaneous whole and remains in unity, but let him see also that everlasting which continues without bound and has its being in becoming, and those varieties of the everlasting which are established between these. For nobody speaking precisely will wish to relate these closely either to true eternity or to time, since time seems to be altogether connected with process and to be a kind of process. But if the flowing time that measures bodily changes, both of substance and other types, is brought about by the eternal principle of time in nature, which is given its being by nature, and if it is this which, as Damascius says, is ever present whole and entire; and if, similarly, the principle of time in the soul that everlastingly pre-exists is numerically the same as that very complete time that he says exists as a simultaneous whole,[30] this doctrine no longer seems to me to

[30] The first soul (*prôtê psukhê*), viz. the World Soul, is measured by the whole of time, see Proclus *El. Th.* proposition 200.

be rebarbative. For there is nothing surprising if the principle of time, like that of process, should be a simultaneous whole both in time and in nature, in the way that simultaneity occurs in them. For the principle of body also pre-exists, incorporeal and indivisible, and that divided body, that cannot be indivisible, proceeds from it. Thus, then, let both time and process be simultaneous wholes in the realm of paradigms and formal principles;[31] but in my opinion they cannot be so in the realm of copies and created existences. But perhaps not even the natural and psychical principles, if they have fallen away from the truly real and what remains in unity, can be simultaneous wholes, except only as is fitting to those intermediates. So, to sum up, I think that it is impossible to regard time as a simultaneous whole when it is in participation, but from the analogy with eternity I have myself also come to the idea of the primary time which is above all things that are in time and which makes them temporal by their participation in itself, that is, which orders and measures the duration of their being and which makes the parts of that duration have relative position.[32] For just as eternity exists prior to eternal things and is a mean between the unified transcendence of being and the separated descent of the intellect, existing as something differentiated (wherefore eternity is coordinate with life also, since life is also between what truly is and intellect, that is, between the unified and the differentiated, and so exists as something differentiated), so, then, between what truly is and what comes to be, that is, the unchanging and the changing, or that which has simultaneously its whole being in essence, power

[31] *logoi*: in Neoplatonism they refer usually to the principles by which things are formed. There is a clear continuity with the Stoic reason-principles. For the Stoics Nature is a universal principle, the exemplar for human nature. But as they considered reason to be the best faculty by which to lead a life in accordance with that nature, they thought that *logoi* are implanted by universal nature in individuals as 'seeds', viz. as seminal reason-principles (*spermatikoi logoi*). Neoplatonists extended it to the scheme of Creation, and Nature as hypostasis. Proclus, for example, uses 'seminal principles' in *in Tim*. 1.143,17-18.

[32] Simplicius' idea of a primary time above the universe (*prôtou khronou tou huper panta*) fits with his general fondness for 'supranatural' causes for the motion of the heavens, etc. For a discussion on the innovations of Damascius' theory of time and comparisons with pseudo-Archytas, Iamblichus, Proclus and Simplicius, see S. Sambursky and S. Pines, *The Concept of Time in Late Neoplatonism*, Jerusalem 1971, 9-21.

and activity, and that which is not a simultaneous whole,[33] there must by all means be that which is directly exempt from becoming, but has descended from being and therefore measures, holds together and orders the duration of becoming by participation in itself. For the participated time is not duration, but the measure and ordering of duration.

If I am right about this, this primary time is related to the soul as unparticipated eternity is to life. For neither is life eternal (for the eternal is that which is measured by eternity), though it has the same nature as the eternal, viewed from a different perspective, nor is the soul in time but is its own time. The only difference is that soul exists as life-creating, but time as the measurement of the duration of being, unless, indeed, procession has divided their natures, so that soul is one thing, time another. For THERE also there must be this threefold intermediacy, in one way viewed as life, in another as eternity and in another as wholeness;[34] but these are not divided in themselves, though we make divisions in their unified totality. It is clear that this must be the time that is honoured as a god by the Chaldeans and other holy religion;[35] but it is not this that natural scientists study but that which is viewed in participation. So that must be enough on this topic.

But let us consider the opinions about time of earlier thinkers. For perhaps some will be found to have touched on the time we have now mentioned. The Pythagorean Archytas, then, who seems to be the first of whom we know by report to distinguish being from time, wrote the following about time in his work *Concerning the Whole*:[36] 'A particular time and

[33] Supplying e.g. *kai tou mê hama holon ekhontos to einai* at 784,29. (J.O.U.)

[34] Wholeness, life and eternity in the intelligibles, derived from Plato *Parm.* 142, and *Tim.* 37-8, see Proclus *Pl. Th.* (Saffrey and Westerink) 3, chs 25, 16, 18.

[35] Time as deity: the Chaldaean Aeon, adjusted by Proclus from the God of Time to that of Eternity, see H. Lewy, *Chaldaean Oracles and Theurgy*, Paris 1978, 99-105; Proclus, *in Tim.* 3.16,5-10; 2.54,10-16. There was also the option of identifying Time Unparticipated with Cronus, the Father 'remaining' in Intellect. It derived from a very old and common substitution of Time, in Greek *Khronos*, with *Kronos*, Zeus' father. It was believed to have been first used philosophically in the sixth century BC by one of Pythagoras' teachers, Pherekydes (for the problems see H. Schibli, *Pherekydes of Syros*, Oxford 1990, 27-38). In late antiquity it was adopted widely and became well-established among the syncretic religions of the Roman Empire, particularly in Mithraism, the favourite religion of the Emperor and the soldiers. See H. Lewy, ibid., 405-6; L.A. Campbell, *Mithraic Iconography and Ideology*, Leiden 1968; F. Cumont, *The Mysteries of Mithra*, New York 1956.

[36] Archytas the pseudo-Pythagorean, first to distinguish time from being, cf.

universal time have the property of being without parts and non-existent. For the now that is without parts, even while we speak and think of it, has passed away and has no permanence. It comes into being continuously, but the numerically identical is never preserved, only the qualitatively identical. For the present now and the coming now are not the same ones as that which preceded them. For this is departed and is no longer, and that is no sooner conceived of and present than it is gone. Thus one now is ever continuously joined on to another and another which comes to be and passes away, though they are qualitatively the same. For every now, simple and indivisible, is the end of the past and the beginning of the future, just as when a line is cut the point round which it is cut becomes the beginning of one line and the end of the other. But time is continuous and not discrete, as are number, words and melody. For syllables are the parts of words, and these are discrete, and sounds are the parts of melody and units of number; but a line, an area and a place are continuous. For the parts of these, when divided, have common boundaries. A line is cut at a point, a plane at a line
786,1 and a solid at a plane. So time also is continuous; for there was never a natural world when there was no time, nor process when there was no now. But nows, one after another, ever were and will be and will never be exhausted; each is numerically different, qualitatively identical. But time differs from other continuous things in that the parts of a line and an area and a place exist, but of time the past parts have passed away and those to come will pass away. So, therefore, time either does not exist at all or it exists faintly and scarcely. For how could that of which what has been no longer exists, what is to be is not yet, and what is now is simple and indivisible truly exist?'

But the divine Iamblichus, in the first book of his

---

Simplicius *in Cat.* (*CAG* 8) 348,20-8; 356,28-36. This 'Archytas' is not the fourth century BC Pythagorean, but a Neopythagorean (first century BC and later). Late antique philosophers, including Neoplatonists, regarded the Pythagoreans and Plato as the original authors of many concepts which Aristotle made famous or plagiarised. Thus 'Archytas' was widely credited as the precursor of Aristotle's categories; 'Ocellus of Lucania' as the precursor of the hot-cold-dry-moist theory, Proclus *in Tim.* 2.38,13-16 (J.M. Dillon, *The Middle Platonists*, 156 n. 1, notes that according to Philo some thought that Ocellus' *Universi Natura* was the 'Pythagorean' source of Aristotle's theory of Elements).

commentaries on the *Categories*,[37] says that Archytas defines time as being 'a number of a certain process, or a universal dimension of the nature of the Whole'.[38] He himself explains the definition: 'a certain process' refers not to one out of many (or the others will be excluded from time), nor to the sum of the many processes (for this is not one), but to that which is truly single and prior to all others, as being the unit of process which is rightly primary and the source of all other processes. It is the first modification of the soul, coming to be through the procession of the principles. The number of this process does not supervene or come from outside, as Aristotle thinks, but is set above the process in the causal order and advances it according to fitting measures, being an essence advancing essential activity and, as it were, bringing to birth the self-moving procession of the essential principles of the soul. 'One must suppose', he says, 'that by "the universal dimension of the nature of the Whole" the Ancients meant the continuum grasped in the principles and divided into parts. For what it displays in the world of becoming – the process of this now from the former now and of that now from the one before it – this is seen as pre-existing much prior and more creatively in regard to the being of the principles of natural wholes. This also fully completes the dimension of the eldest time of all, which holds together the principles of nature. The views of even earlier thinkers', he says, 'are consistent with this. For some of them defined time, in accordance with the derivation of the Greek word, in terms of some dance of the now,[39] others by the periodic changes of the soul, others by the natural

[37] Simplicius is our source for the information that Iamblichus composed a commentary on Aristotle's *Categories*, though most of the refs are, naturally, in Simplicius' own commentary on the *Categories* (e.g. 2,9ff.). The Neoplatonists were frequently commentators on *both* Plato and Aristotle. J.M. Dillon, *Iamblichi Chalcidensis in Platonis Dialogos Commentarium Fragmenta*, Leiden 1973, 21, points out that Iamblichus was the first to use Archytas in the exegesis of the *Categories*.

[38] Reading *kinasios tinos arithmos ê kai katholô* instead of *kinasios tis arithmos ê kai katholô* at 786,12-13. Cf. *kinêseôs tinos* at 786,13 and 787,30. See also *Simplikios: Über die Zeit*, Anhang IV, by Erwin Sonderegger in *Hypomnemata*, Heft 70, Göttingen 1982. (J.O.U.)

[39] Time (*khronos*) defined as the moving instant, the 'dancing now' (*khoreia nun*). For time as a flux of the instant, and the important distinction between intelligible and sensible time, see S. Sambursky and S. Pines, *The Concept of Time in Late Neoplatonism*, Jerusalem 1971, 14-21; R. Sorabji, *TCC*, 33-66.

container of these changes, others by cyclic recurrence, all of
787,1 which were taken over by the Pythagorean sect. And now,' he says, 'by our exegesis we have revised the definitions into two: but we must contract these two accounts into one, and make time out as both continuous and divided, even if continuity is more important.'

That is how Iamblichus says that both psychic and physical time were represented by Archytas. But by the passage of Archytas that I quoted earlier on he says that the activity of time upon the exterior world is revealed, but that Archytas does not merely consider that time exists but also that it holds precedence among beings in accordance with its own well arranged order to which is owed the relation of precedence and subsequence in our actions; this would not have been the case had time not existed antecedently. Iamblichus considers that being simple and being unreal should be taken as applying to different sorts of time, assigning simple time to the forms of principles, that are independently permanent, and assigning the non-existent to the processes that proceed from them, since these do not preserve a simple and unchanging essence. He also assigns the simple time to the activity and perfection that remains unchanged in essences, and non-existent time to the inclination away from being into becoming, since it has not preserved the purity of its first nature. 'And where', he says, 'must we conceive the flux and degeneration of time to be? We shall say that it is in things that participate in it. For these things that ever come to be cannot receive without change the static nature of that other time. But at different times with different parts of themselves they make contact with it and their condition is falsely attributed to it. To come into being at a now will be an attribute of those things which ever participate in a now, whereas the identity in a single continuity of things that are constantly changing is a property of the indivisible now, and from it is imparted to things constantly changing. Therefore the numerical difference, ever changing, is a proof of the mutability of participating things, but the form, remaining identical, exhibits the identity of the indivisible now.' That is how Iamblichus interpreted Archytas, as well as contributing much beyond the exegesis.

But Damascius, I think, if more prosaically,[40] yet more suitably to the text of Archytas, understands the 'number of a certain process' as being not of process as an unchanging form, but of changing process; so that it is not of the psychic alone, but equally of all change. Very likely Archytas says 'a certain process' because alteration is always individual and concerned with certain individuals, whereas the universal is unalterable. It is the universal dimension of the natural Whole because it measures not only process but also rest, which Aristotle well recognised, saying that for process 'this is what it is to be in time, that its being should be measured. It is clear that for other things also this is what it is to be in time, that their being should be measured by time' [*Physics* 221a6ff.]. So in the case of rest, also, its duration is measured by time. So that, even if time is predicated of process, it is predicated of this continuance of being within becoming. Aristotle also seems to conjoin the duration of nature with the continuance of each thing's being.

So, having called time the number of process, Archytas moved to a more general conception of the duration of nature proper to things in becoming, which is observed also in rest and in general in all generated and natural being. That is why he used the expression 'and the universal dimension of the nature of the Whole' since he wished to consider the nature of time particularly in the natural world. For the soul also participates in time, to the extent that it shares in nature and becoming, as well as in eternity, when it reascends into true being. Also 'dimension' is more appropriate to time than is 'number'. For the 'nows' do not stand at intervals from each other as do individuals, since nothing comes between two nows that is not a now. Therefore time is continuous but not discrete, as a sentence is composed of nouns and verbs and syllables and letters, and a melody of notes.

But, since the dimension of size is a different one, he added 'of the nature of the Whole', calling nature that offshoot of

[40] Damascius interprets more prosaically than Iamblichus. Many Athenian Neoplatonists thought that Iamblichus had his head too much in the clouds, despite respecting him as 'divinely' wise. Even Proclus occasionally said that Iamblichus takes a too 'elevated' view of things, *hupsêlologoumenos*, *in Tim.* 1.19,9; 1.426,5. Cf. also Damascius' attitude to Proclus, 795,15. Despite modern opinion, the later Neoplatonists were critical of and quietly irreverent to their authorities.

being which ever comes to be and flows along. Continuing, he made it still clearer that he did not define dimension in terms
20 of size but in terms of everlasting continuity. He calls time 'simple' and 'non-existent' since he sees it as indivisible with regard to the now, but non-existent because it does not persist numerically identical. For such is the nature of things that have their being in becoming. Therefore in these there is process and before and after. He also sees that when he says
25 that time is simple and non-existent, and when he says that the time at a now is different from time that has gone by he is treating 'time' and 'now' as synonyms. But when he calls the now the end point of time gone by and the beginning of the future, the now is something different, and also when he calls the now indivisible, but time continuous, and when he calls time a number; for he would not call the now a number if it is
30 indivisible. So it seems that Archytas also, like Aristotle, treats time as the continuous and inexhaustible flow of nows; above all he teaches that the time that is involved in becoming and is properly so-called is the true image of eternity.

We have already stated the view of Plato and Aristotle about time. Theophrastus and Eudemus, the colleagues of
35 Aristotle, seem to have believed and taught the same view of time as Aristotle. But Strato of Lampsacus criticised the definition of time given by Aristotle and his colleagues,
789,1 although he was himself a pupil of Theophrastus, who followed Aristotle on almost everything, and followed a newer path.[41] For he does not accept that time is the number of change, because a number is a discrete quantity, process and time are continuous, but the continuous is not denumerable.
5 But if, since there are different parts of a process, some earlier and some later, there is for this reason some number of process, then length should have a number in the same way, for here too there are different quantities; similarly in other continuous processes there is a before and an after, so that there would be a time of time for time. Moreover, there is no

[41] = Theophrastus fr. 151B FHS&G, *Theophrastus of Eresos*, eds W.W. Fortenbaugh, P.M. Huby, R.W. Sharples and D. Gutas, Leiden 1991. I am grateful to R.W. Sharples of Project Theophrastus for the full reference. See also E. Zeller, *Hist. Phil.* II 2.911. On the moot subject of Theophrastus' merely following Aristotle, Boethius agrees in *in Int.* (2), ed. Meiser, 12,9-12, but Theophrastus' questioning of Aristotle on place is only one of several reasons for scepticism (R. Sorabji, *MSM*, 199).

coming to be and passing away in number, even if numerable things pass away, but time continuously comes to be and passes away. Further, all the parts of a number must necessarily exist – for there could not be a triad without three integers. But this is impossible in the case of time, for the earlier and the later will have to exist simultaneously. Again, an integer and a now will be the same thing if time is a number, since time is composed of nows and a number of integers.

Strato raises a further difficulty; why should time be the number of the earlier and the later in process rather than in rest? For there is similarly an earlier and a later in rest. But this one is easy to solve on the basis of the foregoing discussion. For Aristotle said that time was the measure of the whole flow of becoming, which is common to both process and its opposite, rest, and of the being of everything in the world of becoming. He called time a number, not because it is simply a number (for he shows that time is continuous, like size and process), but as being recognised through the division in the continuous of earlier and later by those able to perceive time. So the attacks based on number disturb nothing in Aristotle's conception.

But Strato seems to have a good objection in regard to the account of being in time. 'If', he says, 'to be in time is to be enclosed in time, it is clear that nothing perpetual could be in time.' But Aristotle apparently calls that which exists in all time eternal and not temporal, since not enclosed in time. But if everlasting time has its being in everlasting becoming, and if the temporally perpetual is such as the process of the heavens, being everlasting by continuing without bound, perhaps it is true to say that every time taken is enclosed by another time.

Strato made many other objections to the Aristotelian account. He himself treats time as the quantitative in actions. 'For', he says, 'we speak of someone being away from home, or sailing, or being on active service, or fighting a war for a long or a short time, and similarly we speak of his sitting and sleeping and doing nothing for a long[42] or short time. In these

[42] Reading *polun* instead of *pollên* at 789,35. (J.O.U.)

790,1 the great quantity is a great time, the little quantity a little time. For time is the quantity in each of them. That is why some people say of the same man that he is coming slowly, others that he is coming fast, depending on how the quantity in the occasion appears to each party. For we say that that is quick in which the quantity between beginning and end is little and yet a great deal is done. The slow is the contrary, when the quantity is great but little is achieved. That is why', he says, 'there is no fast or slow in rest. For each period of rest is equal in achievement and there is neither great achievement in a little quantity nor little achievement in a great quantity. This', he says, 'is why we speak of more or less time but not of quicker and slower time. For an action and a process are quicker or slower, but the quantity in which the action occurs is neither quicker nor slower, but is greater or less, as is time. Night and day', he says, 'and months and years are not time nor parts of time, but the former are the presence of light or dark, the latter the orbits of the moon and the sun, but time is the quantity in which they occur.'

But if what is done is one thing and the quantity in which it is done another, and this is time, we are told that it is quantity, but what sort of quantity it is is unclear. So from this nobody can get a conception of time that he has not got already. Also it is clear that there is something that is quantity apart from the specific character of process and rest, since we speak of much taking place in a little time, when it is quick, little in much time when it is slow. But what this extra is Strato's account has not made clear. He says: 'The reason why we say that everything is in time is that time is a concomitant of everything that is or becomes.'[43] But we often use the opposite idiom; for we say that the city is in disorder or that the man is in fear or in ecstasy because these are in them.

So much, then, as a short reminder of Strato's opinions and problems about time. Also, all the philosophers we have mentioned were concerned with physical time, even if some of them made mention of the separate, distinct time that is prior in the account of causation.

[43] Closing quote after *ousi* at 790,23. (J.O.U.)

Among the moderns, Plotinus seems to have been the first to investigate the primary time. He says that it is the life of the soul in its change of nature from one way of living to another, showing how this fits in with what he has said about eternity. For he says that eternity is the life centering on what is in a state of being, a life that is all in one, complete and in every way unextended.[44] But it is perhaps better to listen to the intellective sayings of Plotinus himself:[45] 'So if it were said that time was the life of the soul in a change from one way of living to another, would it seem reasonable? Well, if 791,1
eternity is life at rest, identical and unchanging and already without limit, and if time must be the image of eternity, related to it as the whole HERE to that THERE, then we must say that, corresponding to the life THERE, there must be another life homonymously so-called in relation to that power of the soul, and, corresponding to the intellective process, a process of some part of the soul, and, corresponding to self-identity, changelessness and rest, that which does not preserve its identity but acts one way and another, and, corresponding to the undivided and one, an image of the one which is one in continuity, and, corresponding to the unbounded and whole, that which is always moving to the next stage without end, and, corresponding to the simultaneous whole, that which will come to be part by part and will ever be awaiting completion. For that is how it will imitate that which is already whole, simultaneous and unbounded, if it wishes ever to add being in its being. For so it will imitate the being that is THERE. But one must not suppose time to be something outside the soul, just as THERE we must not suppose eternity outside being. Nor is it consequential or later, any more than THERE, but one must suppose it visible in it, internal and allied to it, as is eternity THERE' [*Enneads* 3.7.11]. He adds that it is not time that measures process, but process that measures time, since time is imperceptible but process perceptible; for the imperceptible is recognised and measured by means of the perceptible. So he says that 'what is measured by the circular motion of the heavens (for that is the revealed) will be time, which is not produced by the circuit,

[44] Plotinus *Enn.* 3.7.3
[45] ibid. 3.7.11

but revealed by it' [*Enneads* 3.7.12]. 'This', he says, is why people were led to call time the measure of change, instead of saying that it was measured by change, and then to add what it was that was measured by change' [*Enneads* 3.7.13]. He also adds this: 'Is time in us also? Or is it in every soul of that kind, and of the same form in them all, and are they all one? In that way time will not be split up, since eternity is not, which, in another way, is in all things of the same form' [*Enneads* 3.7.13]. In these passages Plotinus believes that time is the changing life of the soul, because eternity is the life centering on what is in a state of being, a life that is all in one, complete and indivisible, as was said before, and he seems to have got such an idea by analogy. For, as in the intelligible world, after that which comes first into being, which he calls intellect, there is the intellectual life proceeding together with eternity (so that what is after it[46] is both life and eternal), so, also, on the margin of the created world, after the indivisible being, the life of the soul goes on together with time.

Damascius objects to him that in his teaching he substitutes eternal intellect for eternity. But perhaps the life centering on what is in a state of being, if it is that which is immediately after that which is, is not that which belongs to the eternal intellect in third place, but to that between them, which Plotinus also calls intellect, just as he does the first. But I think that this also is worth consideration, whether even among intelligibles life and eternity are unified, while we, making distinctions within what is THERE unified, call that mean state THERE sometimes life, sometimes eternity, sometimes a whole and parts, as we have learnt in the *Parmenides*,[47] but in those things which have joined with becoming the unity in life no longer remains, but proceeds to take on the nature of an image, even if it still retains something withdrawn from image and becoming.

But the life of the soul is an image of the intellectual life, as Plato himself said.[48] So the changing life of the soul would not be simply time, but would be temporal, as the unchanging life

[46] Reading *auto* instead of *autên* at 791,30. (J.O.U.)

[47] Plato *Parmenides* 142B-143A. The *Parmenides* was the most important dialogue in the Neoplatonic curriculum.

[48] Soul image of intellectual life, Plato *Tim.* 36E.

of the intellect is eternal. That basic time is the one that measures the changing life of the soul, if the alteration of the soul is the very first, and all others stem from it. Perhaps this is the point that Aristotle seized on when he said that there would be no time if there were no soul, because there would be nothing that made the count [*Physics* 223a16]. So let time have its basis in the soul, though it is something additional to soul that measures its changing activities. For it will hold together the being and essential activities of the soul between the indivisible nature and that divided in the realm of bodies, as a mean measure between eternity and time and will measure them most justly.

After Plotinus, let Iamblichus also come to us and let him too provide illumination for our thoughts. For he also speaks about the primary and unparticipated time when he expounds the discourse of Archytas in his commentary on the *Categories*.[49] The activity of the simple is not, like the light of a lamp, in constant process. Nor is it perceptible, nor in flux, but it remains in its procession and always is and is always active and never becomes; it is pre-eminent in the unchanging, being uncreated in the numerically identical form, and is indestructible. But he says that the now is always coming to be; whereas I think that it is clear on the same basis that the becoming once began to become and does not become for ever, and that the now is and does not become. But what becomes, having process in its duration, does not become in the now. For it is rather rest that is seen in the now and not process.

Iamblichus says that the simple is to be conceived as continuous and the measure of continuous process, and as the genetic cause of time. Where, then, is the flowing time that comes to be? Presumably in participating things. 'For', he says, 'whatever comes to be cannot receive the unchanging, static being of the time THERE; but sometimes with various portions of themselves they lay hold on that being and their condition falsely mimics that being. So becoming will belong 793,1
to those things that for ever participate in the now, but the unitary and continuous identity in things that are inconstant

[49] For Iamblichus on Unparticipated and Participated Time, see also Proclus *in Tim.* 3.14,16; 3.30-3.

is the property of the indivisible now.' In this passage he seems to postulate a single uncreated now that is prior to participating things, and derived from it the nows attributed to the participants. As in the case of the nows, so there is a single time which precedes temporal things but many times that come to be among participating things, among them the past time, the present time and the future time. 'This is so', he says, 'not through the ordering of our own actions, but by the reality of time that rightly orders our actions. For', he says, 'one could not reckon which action came first and which second if time did not exist in itself, with reference to which time had an order.'[50] That he regards time as encompassing the soul is shown by the following passage:[51] 'Time has been reasonably defined as a proceeding image of eternity, the soul being likened to the intellect, its thoughts assimilated to intellections, and the simple now in the soul being likened to that which rests in the one, and what encompasses things HERE is likened to what THERE simultaneously and ever encompasses realities, and what is in process HERE to what is static THERE, and the measure of coming to be is derived from the measure of essences.'

So he clearly makes eternity the encompassing measure of true realities, and time itself an essence, but one that measures coming to be, the time of the soul first, after this that which proceeds from it, and then remains the time that accompanies process and is not real since it has its being in becoming. He claims that not only the now[52] is present, but also time between two limits. Let us add what he said in his commentary on the *Timaeus*. In Book 8,[53] mainly following Plato, he teaches the close link of time to eternity. Therefore

[50] Closing quote after *praxeôn* at 793,11. (J.O.U.)

[51] For Proclus time was created directly by the Demiurgic Intellect in imitation of Intelligible Eternity, *Pl. Th.* (Saffrey and Westerink) 5.73-4; *in Tim.* 3.27.

[52] R. Sorabji, *TCC*, 40, the 'now' is the *indivisible* instant, as opposed to a stretch of time between two instants (see Introduction).

[53] (On Plato *Tim.* 37D = Iamblichus *in Tim.* fr. 63 (J.M. Dillon, *Iamblichi ... Fragmenta*). Dillon, 60-2, reckons that the reference to an *eighth* book is erroneous: it should be Book 3. He suggests a misreading of the Greek letter γ, *gamma* (numeral 3) for η, *eta* (numeral 8). His two chief reasons are: (i) eight books up to Plato *Tim.* 37D would entail eight exceptionally slim volumes; (ii) it does not conform to the normal exegetical sequence from Plato on time to space, which is said to be in Iamblichus Book 5 (Simplicius 639,24). For a commentary on this Simplicius fr. see Dillon, *Iamblichi ... Fragmenta*, 343-7.

he mainly discusses that time which is separated from the universe, but which encompasses and controls the measures of all the process therein; this presumably is another time beyond that investigated by natural scientists. In Chapter 6 he says: 'We coordinate its actualised essence with the proceeding arrangement of the universe, which is itself coordinated with created things and is inseparable from the things it has brought into being. That the arrangement simultaneously creates the heavens is shown by the fact that the existence of time also is coordinated with the arrangement proceeding from the demiurge. Further, its existence precedes the circling of the heavens, just as the arrangement that arranges it is set immediately prior to it, correspondingly to its superiority. This arrangement all at once confined it 794,1 completely within certain limits, thus preserving the principle according to which it is produced.

'So we, also, agree that there is an ordering of time, but one it imposes, not that is imposed on it; it is not one which follows on things that precede it but is itself the origin of its creations and older than them. Nor is it that which is separated into parts in accordance with principles or processes or other determinate powers, but that which is whole and complete throughout the derivative creation. We do not assign the earlier and later in its order to alterations in processes, or to the unfoldings of life or the passages of cosmic creation, or anything of that sort. Rather, we mark it off according to its initiation of causes and continuous joining together of created things and its fundamental activity and its power of controlling changes and all such things. Furthermore we deny that time and with it the heavens were given birth with the process proceeding from the soul or from life, but say that it came from the intellective ordering proceeding from the demiurge. For with this ordering time and the universe both acquired their existence in him. Further, the old account explicitly explains that the god, in arranging and bringing forth time, created the world with it. So one might hold that it is a measure, not by measuring locomotion, or being measured by process, or revealing or being revealed by the celestial orbit, but by being the cause and unifier of all these things.'

That is what Iamblichus wrote about the time that is

exempt from coming to be and exists in its own right. He also wrote the following in Chapter 10 about this time and that which is given by it to the Universe:[54] 'On this account and according to the pattern of the eternal nature, time also is so far as possible self-identical. It resembles eternity and is so far as possible self-identical because of its simple nature, and on its own, with a single activity, it becomes present and proceeds and bounds all things that come to be in the same way, though they are different.' He also presents another proof as follows:[55] 'The paradigm exists through all eternity, but the universe has come to be through all time without end, so that it is both present and future. What exists as a pattern in the intelligible world exists as an image in the generated world. What is THERE eternal is HERE temporal; and that which in the intelligible world is in being now and present, that in this world in continuity comes into being without end. The unchangingly self-subsistent appears in these regions as past, present and future. What THERE is undivided is HERE seen as divided. And now the intermediate dual nature of time has become clear; it is intermediate between eternity and the heavens, and it is dual insofar as it exists together with and in relation to the universe, but is ordered in relation to eternity; it is set over the one, and is a likeness of the other.' Such is the clear meaning of the relevant passages of the *Timaeus*, according to the divine Iamblichus.

But Proclus, the Lycian philosopher and the guide of our teachers, also holds roughly the same philosophical view about the separated time as Iamblichus, and strives to demonstrate that it is not only intellect but also a god, so that it has even been called on to appear by the theurgists.[56] He says that this time has its internal activities unchangeable, whereas those reaching beyond it are in change. However, concerning the participated time that is inseparable from becoming he maintains the same view as Aristotle, believing that Aristotle says that time exists only in the now. Proclus' successors right up to our time have followed him not only on

[54] (On Plato, *Tim.* 38BC = Iamblichus *in Tim.* fr. 67 (Dillon, *Iamblichi ... Fragmenta*).

[55] = Iamblichus *in Tim.* fr. 68 (Dillon, *Iamblichi ... Fragmenta*).

[56] Proclus on Unmoved and Moved Time, and as deity, see *in Tim.* 3.26-27,30.

this point but in all other matters. I except Asclepiodotus, the best of Proclus' pupils, and our Damascius, of whom the former, because of his extreme cleverness, rejoiced in novel doctrines, while Damascius, through rivalry[57] and his sympathy with Iamblichus, did not hesitate to reject many of Proclus' doctrines.[58] With regard to the opinions of these two philosophers, suffice it for me to say that if those who have sought the cause of time among intellects and gods have said that it, too, is an intellect and unchanging and a god, we must accept it. For if anyone seeks the first causes of process and coming to be he will most certainly find them to be intellect and god. There is nothing surprising if they should call time itself by the same names, since this has often seemed good to theologians, and perhaps also to the gods themselves. But if anyone is enquiring into the generally recognised time which is present in process, I do not think it possible to call it unchanging or existing as a simultaneous whole or intellect just as it is not possible to think of process as unchanging or existing as a simultaneous whole.

But among the problems about time one thing remains for us to do, which is to tackle the difficulties that Aristotle raises about the existence of time, and solve them. For it is clear that if these are not resolved nobody will be firmly confident that time exists at all. The dissolution of these difficulties is particularly worth taking seriously because Aristotle himself, while he finally resolved the difficulties that he raised initially in his discussion of place, left these unresolved. Of his subsequent expositors, none, I think, has given the solutions, not even the most industrious of his expositors, Alexander of Aphrodisias.

So that we can easily recall these difficulties, let us have Aristotle's own words before us: 'On the following grounds one 796,1
might suspect that time does not exist at all, or barely and obscurely. One part of it has been and is not, while the other is going to be and is not yet. But both boundless time and whatever time one takes, is made up of these. But one would think it impossible that what is composed of non-existents

[57] Reading *philopthonian* instead of *philophonian* at 795,16. Possibly *philoponian* as Sambursky and Sonderegger; cf. 795,34. (J.O.U.)

[58] This is a most telling statement on the extent of Proclus' influence.

should itself share in being. Further, if a divisible thing is to exist, it is necessary that, when it exists, all or some of its parts should exist. But of time some parts have been, others have yet to be, but no part of it is present, though it is divisible. For the "now" is not a part; a part has a measure, and a whole must consist of its parts. But time does not seem to be composed of nows. Further, it is not easy to see whether the now, that seems to divide the past and the future, always remains one and the same, or is always a different one. For, if it is ever different, and if no two parts of time ever exist simultaneously, unless one contains and the other is contained, as the shorter is by the longer, then that which now is not but which once was must have passed away sometime, and nows will not be simultaneous with each other, but the earlier will have to have passed away. But the now cannot have passed away in itself, because it then is, but the former now cannot have passed away in another now. For we must accept that no now can adjoin another now, just as no point may adjoin another point. If, then, it has not passed away in the adjoining now, but in another, it will be simultaneous with the limitless number of nows in between; but that is impossible. But, on the other hand, it cannot remain ever the same, for there is no single limit to a divisible limited whole, whether it extends continuously in one or many directions. But the now is a limit, and one may find a limited time. Further, if simultaneity in time, i.e. not being before or after, is to be in the same thing – the now – then, if the earlier and the later are in this now, what happened ten thousand years ago should be simultaneous with what has happened today, and nothing is after or before anything else' [*Physics* 217b32ff.].

The philosopher Damascius attempts to resolve these difficulties, by claiming that the present time should not be understood in terms of the indivisible now. For such a now is the limit of a time and not a time. But if what is bounded does not exist, the limit also will not exist. But no process or alteration will appear in such a now; for how will something divisible exist in something simple? But let the argument that raises the difficulty speak for itself. To quote his very words: 'I am impressed at how people dispose of Zeno's argument, [by

saying] that process does not take place in some indivisible unit, but advances completely by a whole step, and does not 797,1
always require the half before the whole, but sometimes, as it were, leaps over whole and part. Those who said that only the indivisible now exists did not realise that the same point arises in the case of time, since it is ever conjoined with process and, as it were, runs alongside it, so that it advances with it with a whole continuous step[59] and does not go through a limitless series of nows. This is so since process is evident in things, and Aristotle has made it luminously clear that no process or alteration occurs in a now, but that at a now process or alteration has occurred,[60] while the process or alteration wholly takes place in time. At any rate, the leap that takes place in process, which is a part of process, will not be going on in the now, nor will that which is present occur in time that is not present. So that that in which a present process takes place will be a present time, that will be divisible without limit, like the process. For each is continuous, but everything continuous is divisible without limit.'

Having said this he sets beside it the passage of Aristotle in which he solves Zeno's paradox, which runs as follows: 'For there will be the same divisions of time and size. If either is unbounded, so will the other be, and the one in the same way as the other. Thus if time is boundless at its extremities, so with size; if time is divisible without limit, so is size; if time is unlimited in both ways, so is size. So Zeno's argument makes a false assumption that it is not possible for a thing to pass through or to come in contact with limitless things in a limited time. For size and time and in general everything continuous is called unlimited in two ways – in respect of divisibility or in respect of its extremities. So while it is not possible to come in contact with an unlimited quantity of things in a limited time, it is possible in the case of the divisible without limit; and

[59] On the divisible leaps of time, see Introduction and R. Sorabji, *TCC*, 52-60. Time does not proceed by atomic instants but by leaps.

[60] R. Sorabji, *TCC*, 10-12. The perfect tense 'process has occurred' (*kekinêtai*) is vital for understanding Aristotle's 'luminously clear' solution to the paradox of the ceasing present instant (see 799,35-800,16 and Introduction). We cannot say that an instant is ceasing to exist (present tense), but we can say that the present instant *has* ceased to exist.

time itself is without limit in this way. So it comes about that the unlimited is traversed in an unlimited and not a limited time, and the contact is with parts that are without limit, not limited' [*Physics* 233a16ff.].

From the foregoing it is clear that he does not recognise being in actuality within continuous things, but, equally, neither the sudden whole step nor the leap over the whole part seems to me to be reasonable in the case either of process or of time. In the case of place, where the parts persist, I think it is possible to envisage such a collocation of parts, but in the case of things that have their being in becoming it would not be possible to find such being all at once, except in our conception of it. For that being all at once will have to be taken as static, not as flowing, and not as coming to be but as being. But what of this sort is there in things that have their being in becoming? But we should rather pay attention to these sayings of Damascius in his own words:[61] 'Again, the everlasting, that is never gathered into one but has its being in becoming, is that time which is day and night and month and year. For none of these comes all at once, as in a contest which, though present, is completed part by part, or a dance; the dance proceeds part by part, but still a person is said to be dancing a present dance. Similarly, the whole of time is present as coming to be, but not as in being.[62] Again, we call the common properties of the forms eternal, as everlasting kinds, numerically in flux but statically the same in form. In this way we preserve the continuity which from our point of view is divided into three and the time which is present for us. The present time is different for different people, whereas in itself it is one and continuous. All this being well said, we must say that the division of time is potential. And we must say that the simple now is a potential division [of time], but our thought does the dividing, and that this now exists as a boundary and as the simple, while we combine as static certain amounts of time, such as days, months and years, not distinguishing them by a single form. The emanation of these

61 Text seems corrupt at 797,35-36 but the sense is clear. (J.O.U.)

62 Introduction and R. Sorabji, *TCC*, 57, and ch. 1. Damascius exploits examples of human activities to point out that things can always have their being in becoming: the competitive race and the dance exist in the present as long they are accomplished in some part of time.

forms contains a large share of being, but it has its being in becoming. If anyone should wish to have the form static and all at once, he will no longer get this same thing that is coming to be, but that which is separate and removed, as a river exists in this world. For each river is a static form, from which the flowing river exists, getting its form in flux. If you halt the river, the river will no longer exist. In this way, then, the present, the past and the future are in form subsumed under the same form, but are unfolded in becoming, that which is ever coming into being being called the present,[63] that which has perished the past and that which is not yet the future. Like process, the whole of time is continually in flux. By taking away present time as being in actuality bounded by the nows on both sides, and stopping it all at once, one would have destroyed the form of time which has its being in becoming, as does process. So I think that the difficulty arises because the soul tries to understand everything in accordance with the changeless form within it. So thus it brings process to a halt, trying to understand it in its formal nature and not in accordance with its natural flux. Just as the soul divides up the intelligible unity, not being able to draw together its wholeness, but sees justice separately, temperance separately and knowledge separately, though each thing THERE is all; and just as the soul reasons that it is immortal, positing three different terms, soul, self-motion and immortality, although the soul *qua* soul contains both self-motion and immortality within itself; as then it is affected in this way towards the intelligible and unified objects, dividing in itself their unity THERE, and believing that they are themselves such as they are understood by the soul, in the same way I think that because of the changelessness of the forms within it, the soul tries to halt the river of becoming. Having circumscribed three separate portions of time, it halts them at the present, taking what is divided altogether. For being in its essence intermediate between things that become and those that are, it tries to understand each group according to its own nature, dividing up the latter in a way inferior to them but more akin to itself, but bringing together the former in a way superior to

[63] Reading *enestôs* instead of *huphestôs* at 798,22. (J.O.U.)

coming to be but more familiar to itself. So thus it recognises days, months and years, joining each together under one form and circumscribing sections of the whole of flowing time.[64]

If, then, my notion in saying this does not altogether lack foundation, I claim from it to dissolve the puzzles about time. The first of these says: since neither the past nor the future exists, but out of these without bound the time taken to be everlasting is composed, that which is composed of things that do not exist would seem incapable of sharing in being. It is clear that he who sets this puzzle does not take account of the flux of becoming, nor does he divide from simultaneous wholes those things that have their being in becoming, in regard to which one could retort that unless some parts of them have come into being and some are not yet, then either some or all must be in being.[65] For it is not in the nature of these things to have being at all, but to become. This is the mode of their reality, since their form is clearly a flowing whole.

But since time at the now seems to have static being, and since it seems possible to solve the puzzle in this way also, on the ground that it is a part of the divisible time, they show[66] that the now is not a part of time with two objections. (1) Every part is a measure of the whole, so that a part of time is likewise. So if the now that is simple is not a measure of time that is continuous, it is clear that the now is not in every way a part of time. He also gives this proof: (2) A part of time contributes to the whole of time, for a part is that out of which the whole is constituted. But the now does not contribute to time (for time is not composed of nows, as he shows in the treatise *On Process*).[67] Therefore the now is not a part of time.

Let us take this as well said.[68] For time is not static at this now. Nor does the now, which they have defined, exist in actuality, as he who allows that the now which is a limit of

[64] R. Sorabji, *TCC*, 59. For Damascius, time's division into past, present and future, or in days, months and years, is inserted only in thought, it is not actual. Further it is not possible to divide something which is essentially flowing.

[65] sc. which is absurd.

[66] Reading *deiknusi* instead of *deiknuousin* at 799,20; cf. *deiknusi* at 799,24. (J.O.U.)

[67] Simplicius refers to an Aristotelian treatise on motion (*peri kinêseôs*) as a separate work: see 800,23 where he says explicitly that this is the sixth book of *Physics*.

[68] Let us take this as well said. (Sc. it does not make any difference). For …

time has being will have to concede being to that which is limited by it.[69,70] The succeeding puzzles attempt to show that the now is not a non-existent boundary of time, but nor can it, as it were, possess some temporal being, since it is necessary, if it is present, for it either to remain the same or to become another and yet another. But he shows that both these are impossible. They take the now to be supposed as in actuality and to be taken as a part of time, of which neither is true.

Next [*Physics* 218a17],[71] he claims that, if the now passes away, it either passes away in itself or in another now, since what passes away passes away in time, just as what comes to be comes to be in time. It is clear that this account presupposes a time of time. Yet he himself denounced[72] the idea that there is process of process. In general, if we try to find measures of measures we shall go on without limit, taking as the measure of the yard-measure another yard-measure and postulating numbers prior to numbers. If those ideas are absurd, where each thing is able to share its own special nature with others that require it, but does not need to share in the very thing which it itself is, that too is an absurdity.[73] If anyone says that it does so need, he says so, I think, under pressure, and it pleases him that the measure should none the less take a share in itself.

So it is not necessary that time should pass away in time nor the now in the now, but there cannot be several at once. For in the flow of time the occurrence of the now is being seen against the background of some sort of conceived rest. But how, if time has its being in becoming, being itself in process, will it avoid needing a time to measure and order the parts of

69 Omitting *einai* at 799,29 and *kai to peratoumenon* at 729,30. (J.O.U.)

70 It does not matter that the sizeless present instant is not a part of time. The point is that Damascius' present 'leap' *is* a part of time (see Introduction, and R. Sorabji, *TCC*, 56-7). Those who are fond of paradoxes, who concentrate on sizeless present instants, had better not grant them being, or they will have to grant being to the stretches of time between two instants, which would defeat the paradox of the non-existent parts of time (see Introduction).

71 The succeeding dilemma forms the second of Aristotle's two main puzzles (see Introduction, and R. Sorabji, *TCC*, 60-1): at what time does the present instant pass away? Is there a time for the passage of time? But such regresses are rejected by Aristotle himself.

72 Denounced (*apephêne*), or better, reading *apephê*, used to deny, at 800,1-2. (Ed.)

73 That too: i.e. the postulation of a time of time is as absurd as the postulation of a process of process, or measure of measure.

time, so that they do not pile up on each other? Surely time is in process in this way, as accompanying process as the measure of process. For even the yard-stick is distinct from what it measures, remaining in its true nature as a measure and not needing anything to measure it.

From these solutions it is also possible to solve the difficulties of Strato about the non-existence of time and also those, disseminated by Aristotle, which treat the now also as an actuality. But let him who is not satisfied with this as a solution of those we have mentioned get hold of the work *On Time* by the philosopher Damascius.

Why, then, did Aristotle not solve the puzzles that he had raised? Perhaps because the argument needed the simplicity of the now and the non-existence in actuality of the now, which he proves in the sixth book of the work in the discussion of process. Here he teaches much about the relation of time and process and provides good insights.

# Bibliography

## Place

Myles Burnyeat, 'The sceptic in his place and time', in Richard Rorty, J.B. Schneewind, Quentin Skinner, eds, *Philosophy in History*, Cambridge 1984.

E.R. Dodds, *Proclus, the Elements of Theology*, Oxford 1963.

Pierre Duhem, *Le système du monde: Histoire des doctrines cosmologiques de Platon à Copernic*, 10 vols, Paris 1913-1959 (see esp. vol. 1, chs 4 and 5 for Antiquity).

David Furley and Christian Wildberg, trans., *Philoponus: Corollaries on Place and Void with Simplicius: Against Philoponus on the Eternity of the World*, London 1991.

Edward Grant, *Much Ado About Nothing: theories of space and vacuum from the Middle Ages to the Scientific Revolution*, Cambridge 1981.

Philippe Hoffmann, 'Simplicius: Corollarium de loco', in *L'astronomie dans l'antiquité classique. Actes du colloque tenu à l'Université de Toulouse-le-Mirail, 1977*, Paris 1979, 143-63.

Edward Hussey, *Aristotle's Physics, Books III and IV*, Oxford 1983.

James G. Lennox, 'Theophrastus on the limits of teleology', in W. Fortenbaugh, ed., *Theophrastus of Eresus: on his life and work*, Rutgers University Studies in Classical Humanities 2, New Brunswick and Oxford 1985.

G.E.R. Lloyd, 'Right and left in Greek philosophy', *Journal of Hellenic Studies* 82, 1962, 56-66.

Concetta Luna, 'La relation chez Simplicius', in I. Hadot, ed., *Simplicius: sa vie, son oeuvre, sa survie*, Peripatoi vol. 15, Berlin 1987.

Henry Mendell, 'Topoi on topos: the development of Aristotle's concept of place', *Phronesis* 32, 1987, 206-31.

G.E.L. Owen, *Logic, Science and Dialectic: collected papers in Greek philosophy*, London 1986, v. index s.v. 'Location'.

Marlein van Raalte, 'The idea of the cosmos as an organic whole in Theophrastus' Metaphysics', in W. Fortenbaugh and R.W. Sharples, eds, *Theophrastus as a Natural Scientist*, Rutgers Studies in Classical Humanities 3, New Brunswick and London 1987, with contrasting assessments by Glenn Most and John Ellis.

S. Sambursky, *The Physical World of Late Antiquity*, London 1962, ch. 1.

—, *The Concept of Place in Late Neoplatonism*, Jerusalem 1982 (introduction and texts with translation).

L. Schrenk, 'Proclus on space and light', *Ancient Philosophy* 9, 1989, 87-94.

—, 'Proclus on corporeal space', *Archiv für Geschichte der Philosophie* (forthcoming).

L. Siorvanes, 'Proclus on the Elements and the celestial bodies: physical thought in late Neoplatonism', PhD diss. London 1987.

Friedrich Solmsen, *Aristotle's System of the Physical World*, Ithaca N.Y. 1960, ch. 12.

Richard Sorabji, *Matter, Space and Motion: theories in antiquity and their sequel*, London and Ithaca N.Y. 1988.

Gérard Verbeke, 'Ort und Raum nach Aristoteles und Simplikios, eine philosophische Topologie', in J. Irmscher and R. Miller, eds, *Aristoteles als Wissenschaftstheoretiker*, Berlin 1983, 113-22.

C. Wildberg, *John Philoponus' Criticism of Aristotle's Theory of Ether*, Berlin and New York, 1988.

Wolfgang Wieland, 'Zur Raumtheorie des Johannes Philoponus', in *Festschrift J. Klein*, Gottingen 1967, 114-35.

## Time

Julia Annas, 'Aristotle, number and time', *Philosophical Quarterly* 25, 1975, 97-113.

A.H. Armstrong, 'Eternity, life and movement in Plotinus' accounts of *nous*', *Le Néoplatonisme* (Report of the International Conference held at Royamont, 9-13 June 1969), Paris 1971, 67-74.

W. Beierwaltes, *Plotin über Ewigkeit und Zeit*, Frankfurt 1967 (text of *Enneads* 3.7, with German translation, introduction and commentary).

—, *Proklos. Grundzüge seiner Metaphysik*, 2nd edition, Frankfurt 1979.

Rémi Braque, *Du temps chez Platon et Aristote*, Paris 1982.

P.F. Conen, *Die Zeittheorie des Aristoteles*, Munich 1964.

Pierre Duhem, *Le système du monde*, vol. 1, repr. Paris 1954; vol. 2, repr. Paris 1974 (1st ed. Paris 1913-17).

Marie-Claire Galperine, 'Le temps intégral selon Damascius', *Les Études Philosophiques*, 1980, 325-41.

Philippe Hoffmann, 'Jamblique exégète, du Pythagoricien Archytas: trois originalités d'une doctrine du temps', *Les Études Philosophiques*, 1980, 307-23.

—, 'Paratasis', *Revue des Études Grecques* 96, 1983, 1-26.

Pamela Huby, 'An excerpt from Boethus of Sidon's commentary on the *Categories*?', *Classical Quarterly* 31, 1981, 398-409.

Edward Hussey, *Aristotle's Physics, Books III and IV*, Oxford 1983.

William Kneale, 'Time and eternity in theology', *Proceedings of the Aristotelian Society* 61, 1960-1, 87-108.

W. von Leyden, 'Time, number and eternity in Plato and Aristotle', *Philosophical Quarterly* 14, 1964, 35-52.

H. Meyer, *Das Corollarium de Tempore des Simplikios und die Aporien des Aristoteles zur Zeit*, Meisenheim am Glan 1969.

W. O'Neill, 'Time and eternity in Proclus', *Phronesis* 7, 1962, 161-5.

G.E.L. Owen, 'Aristotle on time', in Peter Machamer and Robert Turnbull, eds, *Motion and Time, Space and Matter: interrelations in the history and philosophy of science*, Ohio State University 1976, repr. in Jonathan Barnes, Malcolm Schofield, Richard Sorabji, eds, *Articles on Aristotle*, vol. 3, London 1979.

S. Sambursky, *The Physical World of Late Antiquity*, London 1962, ch. 1.

S. Sambursky and S. Pines, *The Concept of Time in Late Neoplatonism*, Jerusalem 1971.

Lucas Siorvanes, *Proclus*, London (forthcoming).
Robert Sharples, in collaboration with F.W. Zimmermann, 'Alexander of Aphrodisias, *On Time*', *Phronesis* 1982.
E. Sonderegger, *Simplikios zur Zeit*, *Hypomnemata* vol. 70, Göttingen 1982.
Richard Sorabji, *Time, Creation and the Continuum: theories in antiquity and the early middle ages*, London and Ithaca N.Y. 1983.
Eleonore Stump and Norman Kretzmann, 'Eternity', *Journal of Philosophy* 78, 1981, 429-58.
T.A. Szlezák, *Pseudo-Archytas über die Kategorien*, Berlin 1972.
John Whittaker, 'Ammonius on the Delphic E', *Classical Quarterly*, n.s. 19, 1969, 185-92.
—, *God, Time, Being*, Two Studies, *Symbolae Osloenses*, supp. vol. 23, 1971, 1-66. Second study, pp. 33-66.
H.A. Wolfson, *Philo*, vols 1 and 2, Cambridge, Mass. 1947.

# Appendix

# The Commentators*

The 15,000 pages of the Ancient Greek Commentaries on Aristotle are the largest corpus of Ancient Greek philosophy that has not been translated into English or other modern European languages. The standard edition (*Commentaria in Aristotelem Graeca*, or *CAG*) was produced by Hermann Diels as general editor under the auspices of the Prussian Academy in Berlin. Arrangements have now been made to translate at least a large proportion of this corpus, along with some other Greek and Latin commentaries not included in the Berlin edition, and some closely related non-commentary works by the commentators.

The works are not just commentaries on Aristotle, although they are invaluable in that capacity too. One of the ways of doing philosophy between A.D. 200 and 600, when the most important items were produced, was by writing commentaries. The works therefore represent the thought of the Peripatetic and Neoplatonist schools, as well as expounding Aristotle. Furthermore, they embed fragments from all periods of Ancient Greek philosophical thought: this is how many of the Presocratic fragments were assembled, for example. Thus they provide a panorama of every period of Ancient Greek philosophy.

The philosophy of the period from A.D. 200 to 600 has not yet been intensively explored by philosophers in English-speaking countries, yet it is full of interest for physics, metaphysics, logic, psychology, ethics and religion. The contrast with the study of the Presocratics is striking. Initially the incomplete Presocratic fragments might well have seemed less promising, but their interest is now widely known, thanks to the philological and philosophical effort that has been concentrated upon them. The incomparably vaster corpus which preserved so many of those fragments offers at least as much interest, but is still relatively little known.

The commentaries represent a missing link in the history of philosophy: the Latin-speaking Middle Ages obtained their knowledge of Aristotle at least partly through the medium of the commentaries. Without an appreciation of this, mediaeval interpretations of Aristotle will not be understood. Again, the ancient commentaries are the unsuspected source of ideas which have been thought, wrongly, to originate in the later mediaeval

* Reprinted from the Editor's General Introduction to the series in Christian Wildberg, *Philoponus Against Aristotle on the Eternity of the World*, London and Ithaca N.Y., 1987.

period. It has been supposed, for example, that Bonaventure in the thirteenth century invented the ingenious arguments based on the concept of infinity which attempt to prove the Christian view that the universe had a beginning. In fact, Bonaventure is merely repeating arguments devised by the commentator Philoponus 700 years earlier and preserved in the meantime by the Arabs. Bonaventure even uses Philoponus' original examples. Again, the introduction of impetus theory into dynamics, which has been called a scientific revolution, has been held to be an independent invention of the Latin West, even if it was earlier discovered by the Arabs or their predecessors. But recent work has traced a plausible route by which it could have passed from Philoponus, via the Arabs, to the West.

The new availability of the commentaries in the sixteenth century, thanks to printing and to fresh Latin translations, helped to fuel the Renaissance break from Aristotelian science. For the commentators record not only Aristotle's theories, but also rival ones, while Philoponus as a Christian devises rival theories of his own and accordingly is mentioned in Galileo's early works more frequently than Plato.[1]

It is not only for their philosophy that the works are of interest. Historians will find information about the history of schools, their methods of teaching and writing and the practices of an oral tradition.[2] Linguists will find the indexes and translations an aid for studying the development of word meanings, almost wholly uncharted in Liddell and Scott's *Lexicon*, and for checking shifts in grammatical usage.

Given the wide range of interests to which the volumes will appeal, the aim is to produce readable translations, and to avoid so far as possible presupposing any knowledge of Greek. Footnotes will explain points of meaning, give cross-references to other works, and suggest alternative interpretations of the text where the translator does not have a clear preference. The introduction to each volume will include an explanation why the work was chosen for translation: none will be chosen simply because it is there. Two of the Greek texts are currently being re-edited –

[1] See Fritz Zimmermann, 'Philoponus' impetus theory in the Arabic tradition'; Charles Schmitt, 'Philoponus' commentary on Aristotle's *Physics* in the sixteenth century', and Richard Sorabji, 'John Philoponus', in Richard Sorabji (ed.), *Philoponus and the Rejection of Aristotelian Science* (London and Ithaca, N.Y. 1987).

[2] See e.g. Karl Praechter, 'Die griechischen Aristoteleskommentare', *Byzantinische Zeitschrift* 18 (1909), 516-38 (translated into English in R. Sorabji (ed.), *Aristotle Transformed: the ancient commentators and their influence* (London and Ithaca, N.Y. 1990)); M. Plezia, *de Commentariis Isagogicis* (Cracow 1947); M. Richard, *'Apo Phônês', Byzantion* 20 (1950), 191-222; É. Evrard, *L'Ecole d'Olympiodore et la composition du commentaire à la physique de Jean Philopon*, Diss. (Liège 1957); L.G. Westerink, *Anonymous Prolegomena to Platonic Philosophy* (Amsterdam 1962) (new revised edition, translated into French, Collection Budé; part of the revised introduction, in English, is included in *Aristotle Transformed*); A.-J. Festugière, 'Modes de composition des commentaires de Proclus', *Museum Helveticum* 20 (1963), 77-100, repr. in his *Études* (1971), 551-74; P. Hadot, 'Les divisions des parties de la philosophie dans l'antiquité', *Museum Helveticum* 36 (1979), 201-23; I. Hadot, 'La division néoplatonicienne des écrits d'Aristote', in J. Wiesner (ed.), *Aristoteles Werk und Wirkung* (Paul Moraux gewidmet), vol. 2 (Berlin 1986); I. Hadot, 'Les introductions aux commentaires exégétiques chez les auteurs néoplatoniciens et les auteurs chrétiens', in M. Tardieu (ed.), *Les règles de l'interprétation* (Paris 1987), 99-119. These topics are treated, and a bibliography supplied, in *Aristotle Transformed*.

those of Simplicius *in Physica* and *in de Caelo* – and new readings will be exploited by translators as they become available. Each volume will also contain a list of proposed emendations to the standard text. Indexes will be of more uniform extent as between volumes than is the case with the Berlin edition, and there will be three of them: an English-Greek glossary, a Greek-English index, and a subject index.

The commentaries fall into three main groups. The first group is by authors in the Aristotelian tradition up to the fourth century A.D. This includes the earliest extant commentary, that by Aspasius in the first half of the second century A.D. on the *Nicomachean Ethics*. The anonymous commentary on Books 2, 3, 4 and 5 of the *Nicomachean Ethics*, in *CAG* vol. 20, is derived from Adrastus, a generation later.[3] The commentaries by Alexander of Aphrodisias (appointed to his chair between A.D. 198 and 209) represent the fullest flowering of the Aristotelian tradition. To his successors Alexander was The Commentator *par excellence*. To give but one example (not from a commentary) of his skill at defending and elaborating Aristotle's views, one might refer to his defence of Aristotle's claim that space is finite against the objection that an edge of space is conceptually problematic.[4] Themistius (*fl.* late 340s to 384 or 385) saw himself as the inventor of paraphrase, wrongly thinking that the job of commentary was completed.[5] In fact, the Neoplatonists were to introduce new dimensions into commentary. Themistius' own relation to the Neoplatonist as opposed to the Aristotelian tradition is a matter of controversy,[6] but it would be agreed that his commentaries show far less bias than the full-blown Neoplatonist ones. They are also far more informative than the designation 'paraphrase' might suggest, and it has been estimated that Philoponus' *Physics* commentary draws silently on Themistius six hundred times.[7] The pseudo-Alexandrian commentary on *Metaphysics* 6–14, of unknown authorship, has been placed by some in the same group of commentaries as being earlier than the fifth century.[8]

[3] Anthony Kenny, *The Aristotelian Ethics* (Oxford 1978), 37, n.3; Paul Moraux, *Der Aristotelismus bei den Griechen*, vol. 2 (Berlin 1984), 323-30.

[4] Alexander, *Quaestiones* 3.12, discussed in my *Matter, Space and Motion* (London and Ithaca, N.Y. 1988). For Alexander see R.W. Sharples, 'Alexander of Aphrodisias: scholasticism and innovation', in W. Haase (ed.), *Aufstieg und Niedergang der römischen Welt*, part 2 *Principat*, vol. 36.2, *Philosophie und Wissenschaften* (1987).

[5] Themistius *in An. Post.* 1,2-12. See H.J. Blumenthal, 'Photius on Themistius (Cod.74): did Themistius write commentaries on Aristotle?', *Hermes* 107 (1979), 168-82.

[6] For different views, see H.J. Blumenthal, 'Themistius, the last Peripatetic commentator on Aristotle?', in Glen W. Bowersock, Walter Burkert, Michael C.J. Putnam, *Arktouros*, Hellenic Studies Presented to Bernard M.W. Knox (Berlin and N.Y., 1979), 391-400; E.P. Mahoney, 'Themistius and the agent intellect in James of Viterbo and other thirteenth-century philosophers: (Saint Thomas Aquinas, Siger of Brabant and Henry Bate)', *Augustiniana* 23 (1973), 422-67, at 428-31; id., 'Neoplatonism, the Greek commentators and Renaissance Aristotelianism', in D.J. O'Meara (ed.), *Neoplatonism and Christian Thought* (Albany N.Y. 1982), 169-77 and 264-82, esp. n. 1, 264-6; Robert Todd, introduction to translation of Themistius *in DA* 3.4-8, in *Two Greek Aristotelian Commentators on the Intellect*, trans. Frederick M. Schroeder and Robert B. Todd (Toronto 1990).

[7] H. Vitelli, *CAG* 17, p. 992, s.v. Themistius.

[8] The similarities to Syrianus (died *c*.437) have suggested to some that it predates Syrianus (most recently Leonardo Tarán, review of Paul Moraux, *Der Aristotelismus*,

By far the largest group of extant commentaries is that of the Neoplatonists up to the sixth century A.D. Nearly all the major Neoplatonists, apart from Plotinus (the founder of Neoplatonism), wrote commentaries on Aristotle, although those of Iamblichus (*c.* 250 – *c.* 325) survive only in fragments, and those of three Athenians, Plutarchus (died 432), his pupil Proclus (410 – 485) and the Athenian Damascius (*c.* 462 – after 538), are lost.[9] As a result of these losses, most of the extant Neoplatonist commentaries come from the late fifth and the sixth centuries and a good proportion from Alexandria. There are commentaries by Plotinus' disciple and editor Porphyry (232 – 309), by Iamblichus' pupil Dexippus (*c.* 330), by Proclus' teacher Syrianus (died *c.* 437), by Proclus' pupil Ammonius (435/445 – 517/526), by Ammonius' three pupils Philoponus (*c.* 490 to 570s), Simplicius (wrote after 532, probably after 538) and Asclepius (sixth century), by Ammonius' next but one successor Olympiodorus (495/505 – after 565), by Elias (*fl.* 541?), by David (second half of the sixth century, or beginning of the seventh) and by Stephanus (took the chair in Constantinople *c.* 610). Further, a commentary on the *Nicomachean Ethics* has been ascribed to Heliodorus of Prusa, an unknown pre-fourteenth-century figure, and there is a commentary by Simplicius' colleague Priscian of Lydia on Aristotle's successor Theophrastus. Of these commentators some of the last were Christians (Philoponus, Elias, David and Stephanus), but they were Christians writing in the Neoplatonist tradition, as was also Boethius who produced a number of commentaries in Latin before his death in 525 or 526.

The third group comes from a much later period in Byzantium. The Berlin edition includes only three out of more than a dozen commentators described in Hunger's *Byzantinisches Handbuch*.[10] The two most important are Eustratius (1050/1060 – *c.* 1120), and Michael of Ephesus. It has been suggested that these two belong to a circle organised by the princess Anna Comnena in the twelfth century, and accordingly the completion of Michael's commentaries has been redated from 1040 to 1138.[11] His commentaries include areas where gaps had been left. Not all of these gap-fillers are extant, but we have commentaries on the neglected biological works, on the *Sophistici Elenchi*, and a small fragment of one on the *Politics*. The lost *Rhetoric* commentary had a few antecedents, but the *Rhetoric* too had been comparatively neglected. Another product of this

---

vol. 1, in *Gnomon* 46 (1981), 721-50 at 750), to others that it draws on him (most recently P. Thillet, in the Budé edition of Alexander *de Fato*, p. lvii). Praechter ascribed it to Michael of Ephesus (eleventh or twelfth century), in his review of *CAG* 22.2, in *Göttingische Gelehrte Anzeiger* 168 (1906), 861-907.

[9] The Iamblichus fragments are collected in Greek by Bent Dalsgaard Larsen, *Jamblique de Chalcis, Exégète et Philosophe* (Aarhus 1972), vol.2. Most are taken from Simplicius, and will accordingly be translated in due course. The evidence on Damascius' commentaries is given in L.G. Westerink, *The Greek Commentaries on Plato's Phaedo*, vol.2., Damascius (Amsterdam 1977), 11-12; on Proclus' in L.G. Westerink, *Anonymous Prolegomena to Platonic Philosophy* (Amsterdam 1962), xii, n.22; on Plutarchus' in H.M. Blumenthal, 'Neoplatonic elements in the de Anima commentaries', *Phronesis* 21 (1976), 75.

[10] Herbert Hunger, *Die hochsprachliche profane Literatur der Byzantiner*, vol.1 (= *Byzantinisches Handbuch*, part 5, vol.1) (Munich 1978), 25-41. See also B.N. Tatakis, *La Philosophie Byzantine* (Paris 1949).

[11] R. Browning, 'An unpublished funeral oration on Anna Comnena', *Proceedings of the Cambridge Philological Society* n.s. 8 (1962), 1-12, esp. 6-7.

period may have been the composite commentary on the *Nicomachean Ethics* (*CAG* 20) by various hands, including Eustratius and Michael, along with some earlier commentators, and an improvisation for Book 7. Whereas Michael follows Alexander and the conventional Aristotelian tradition, Eustratius' commentary introduces Platonist, Christian and anti-Islamic elements.[12]

The composite commentary was to be translated into Latin in the next century by Robert Grosseteste in England. But Latin translations of various logical commentaries were made from the Greek still earlier by James of Venice (*fl. c.* 1130), a contemporary of Michael of Ephesus, who may have known him in Constantinople. And later in that century other commentaries and works by commentators were being translated from Arabic versions by Gerard of Cremona (died 1187).[13] So the twelfth century resumed the transmission which had been interrupted at Boethius' death in the sixth century.

The Neoplatonist commentaries of the main group were initiated by Porphyry. His master Plotinus had discussed Aristotle, but in a very independent way, devoting three whole treatises (*Enneads* 6.1–3) to attacking Aristotle's classification of the things in the universe into categories. These categories took no account of Plato's world of Ideas, were inferior to Plato's classifications in the *Sophist* and could anyhow be collapsed, some of them into others. Porphyry replied that Aristotle's categories could apply perfectly well to the world of intelligibles and he took them as in general defensible.[14] He wrote two commentaries on the *Categories*, one lost, and an introduction to it, the *Isagôgê*, as well as commentaries, now lost, on a number of other Aristotelian works. This proved decisive in making Aristotle a necessary subject for Neoplatonist lectures and commentary. Proclus, who was an exceptionally quick student, is said to have taken two years over his Aristotle studies, which were called

[12] R. Browning, op. cit. H.D.P. Mercken, *The Greek Commentaries of the Nicomachean Ethics of Aristotle in the Latin Translation of Grosseteste, Corpus Latinum Commentariorum in Aristotelem Graecorum* VI 1 (Leiden 1973), ch.1, 'The compilation of Greek commentaries on Aristotle's Nicomachean Ethics'. Sten Ebbesen, 'Anonymi Aurelianensis I Commentarium in *Sophisticos Elenchos', Cahiers de l'Institut Moyen Age Grecque et Latin* 34 (1979), 'Boethius, Jacobus Veneticus, Michael Ephesius and "Alexander" ', pp. v-xiii; id., *Commentators and Commentaries on Aristotle's Sophistici Elenchi*, 3 parts, *Corpus Latinum Commentariorum in Aristotelem Graecorum*, vol. 7 (Leiden 1981); A. Preus, *Aristotle and Michael of Ephesus on the Movement and Progression of Animals* (Hildesheim 1981), introduction.

[13] For Grosseteste, see Mercken as in n. 12. For James of Venice, see Ebbesen as in n. 12, and L. Minio-Paluello, 'Jacobus Veneticus Grecus', *Traditio* 8 (1952), 265-304; id., 'Giacomo Veneto e l'Aristotelismo Latino', in Pertusi (ed.), *Venezia e l'Oriente fra tardo Medioevo e Rinascimento* (Florence 1966), 53-74, both reprinted in his *Opuscula* (1972). For Gerard of Cremona, see M. Steinschneider, *Die europäischen Übersetzungen aus dem arabischen bis Mitte des 17. Jahrhunderts* (repr. Graz 1956); E. Gilson, *History of Christian Philosophy in the Middle Ages* (London 1955), 235-6 and more generally 181-246. For the translators in general, see Bernard G. Dod, 'Aristoteles Latinus', in N. Kretzmann, A. Kenny, J. Pinborg (eds). *The Cambridge History of Latin Medieval Philosophy* (Cambridge 1982).

[14] See P. Hadot, 'L'harmonie des philosophies de Plotin et d'Aristote selon Porphyre dans le commentaire de Dexippe sur les Catégories', in *Plotino e il neoplatonismo in Oriente e in Occidente* (Rome 1974), 31-47; A.C. Lloyd, 'Neoplatonic logic and Aristotelian logic', *Phronesis* 1 (1955-6), 58-79 and 146-60.

the Lesser Mysteries, and which preceded the Greater Mysteries of Plato.[15] By the time of Ammonius, the commentaries reflect a teaching curriculum which begins with Porphyry's *Isagôgê* and Aristotle's *Categories*, and is explicitly said to have as its final goal a (mystical) ascent to the supreme Neoplatonist deity, the One.[16] The curriculum would have progressed from Aristotle to Plato, and would have culminated in Plato's *Timaeus* and *Parmenides*. The latter was read as being about the One, and both works were established in this place in the curriculum at least by the time of Iamblichus, if not earlier.[17]

Before Porphyry, it had been undecided how far a Platonist should accept Aristotle's scheme of categories. But now the proposition began to gain force that there was a harmony between Plato and Aristotle on most things.[18] Not for the only time in the history of philosophy, a perfectly crazy proposition proved philosophically fruitful. The views of Plato and of Aristotle had both to be transmuted into a new Neoplatonist philosophy in order to exhibit the supposed harmony. Iamblichus denied that Aristotle contradicted Plato on the theory of Ideas.[19] This was too much for Syrianus and his pupil Proclus. While accepting harmony in many areas,[20] they could see that there was disagreement on this issue and also on the issue of whether God was causally responsible for the existence of the ordered physical cosmos, which Aristotle denied. But even on these issues, Proclus' pupil Ammonius was to claim harmony, and, though the debate was not clear cut,[21] his claim was on the whole to prevail. Aristotle, he maintained, accepted Plato's Ideas,[22] at least in the form of principles (*logoi*) in the divine intellect, and these principles were in turn causally responsible for the beginningless existence of the physical universe. Ammonius wrote a whole book to show that

[15] Marinus, *Life of Proclus* ch.13, 157,41 (Boissonade).

[16] The introductions to the *Isagôgê* by Ammonius, Elias and David, and to the *Categories* by Ammonius, Simplicius, Philoponus, Olympiodorus and Elias are discussed by L.G. Westerink, *Anonymous Prolegomena* and I. Hadot, 'Les Introductions', see n. 2. above.

[17] Proclus *in Alcibiadem 1* p.11 (Creuzer); Westerink, *Anonymous Prolegomena*, ch. 26, 12f. For the Neoplatonist curriculum see Westerink, Festugière, P. Hadot and I. Hadot in n. 2.

[18] See e.g. P. Hadot (1974), as in n. 14 above; H.J. Blumenthal, 'Neoplatonic elements in the de Anima commentaries', *Phronesis* 21 (1976), 64-87; H.A. Davidson, 'The principle that a finite body can contain only finite power', in S. Stein and R. Loewe (eds), *Studies in Jewish Religious and Intellectual History presented to A. Altmann* (Alabama 1979), 75-92; Carlos Steel, 'Proclus et Aristote', Proceedings of the Congrès Proclus held in Paris 1985, J. Pépin and H.D. Saffrey (eds), *Proclus, lecteur et interprète des anciens* (Paris 1987), 213-25; Koenraad Verrycken, *God en Wereld in de Wijsbegeerte van Ioannes Philoponus*, Ph.D. Diss. (Louvain 1985).

[19] Iamblichus ap. Elian *in Cat.* 123,1-3.

[20] Syrianus *in Metaph.* 80,4-7; Proclus *in Tim.* 1.6,21-7,16.

[21] Asclepius sometimes accepts Syranius' interpretation (*in Metaph.* 433,9-436,6); which is, however, qualified, since Syrianus thinks Aristotle is really committed willy-nilly to much of Plato's view (*in Metaph.* 117,25-118,11; ap. Asclepium *in Metaph.* 433,16; 450,22); Philoponus repents of his early claim that Plato is not the target of Aristotle's attack, and accepts that Plato is rightly attacked for treating ideas as independent entities outside the divine Intellect (*in DA* 37,18-31; *in Phys.* 225,4-226,11; *contra Procl.* 26,24-32,13; *in An. Post.* 242,14–243,25).

[22] Asclepius *in Metaph* from the voice of (i.e. from the lectures of ) Ammonius 69,17-21; 71,28; cf. Zacharias *Ammonius, Patrologia Graeca* vol. 85, col. 952 (Colonna).

Aristotle's God was thus an efficient cause, and though the book is lost, some of its principal arguments are preserved by Simplicius.[23] This tradition helped to make it possible for Aquinas to claim Aristotle's God as a Creator, albeit not in the sense of giving the universe a beginning, but in the sense of being causally responsible for its beginningless existence.[24] Thus what started as a desire to harmonise Aristotle with Plato finished by making Aristotle safe for Christianity. In Simplicius, who goes further than anyone,[25] it is a formally stated duty of the commentator to display the harmony of Plato and Aristotle in most things.[26] Philoponus, who with his independent mind had thought better of his earlier belief in harmony, is castigated by Simplicius for neglecting this duty.[27]

The idea of harmony was extended beyond Plato and Aristotle to Plato and the Presocratics. Plato's pupils Speusippus and Xenocrates saw Plato as being in the Pythagorean tradition.[28] From the third to first centuries B.C., pseudo-Pythagorean writings present Platonic and Aristotelian doctrines as if they were the ideas of Pythagoras and his pupils,[29] and these forgeries were later taken by the Neoplatonists as genuine. Plotinus saw the Presocratics as precursors of his own views,[30] but Iamblichus went far beyond him by writing ten volumes on Pythagorean philosophy.[31] Thereafter Proclus sought to unify the whole of Greek philosophy by presenting it as a continuous clarification of divine revelation,[32] and Simplicius argued for the same general unity in order to rebut Christian charges of contradictions in pagan philosophy.[33]

Later Neoplatonist commentaries tend to reflect their origin in a teaching curriculum:[34] from the time of Philoponus, the discussion is often divided up into lectures, which are subdivided into studies of doctrine and of text. A general account of Aristotle's philosophy is prefixed to the *Categories* commentaries and divided, according to a formula of Proclus,[35] into ten questions. It is here that commentators explain the eventual purpose of studying Aristotle (ascent to the One) and state (if they do) the requirement of displaying the harmony of Plato and Aristotle. After the ten-point introduction to Aristotle, the *Categories* is given a six-point introduction, whose antecedents go back earlier than Neoplatonism, and which requires

23 Simplicius *in Phys.* 1361,11-1363,12. See H.A. Davidson; Carlos Steel; Koenraad Verrycken in n.18 above.

24 See Richard Sorabji, *Matter, Space and Motion* (London and Ithaca N.Y. 1988), ch. 15.

25 See e.g. H.J. Blumenthal in n. 18 above.

26 Simplicius *in Cat.* 7,23-32.

27 Simplicius *in Cael.* 84,11-14; 159,2-9. On Philoponus' *volte face* see n. 21 above.

28 See e.g. Walter Burkert, *Weisheit und Wissenschaft* (Nürnberg 1962), translated as *Lore and Science in Ancient Pythagoreanism* (Cambridge Mass. 1972), 83-96.

29 See Holger Thesleff, *An Introduction to the Pythagorean writings of the Hellenistic Period* (Åbo 1961); Thomas Alexander Szlezák, *Pseudo-Archytas über die Kategorien*, Peripatoi vol. 4 (Berlin and New York 1972).

30 Plotinus e.g. 4.8.1; 5.1.8 (10-27); 5.1.9.

31 See Dominic O'Meara, *Pythagoras Revived: Mathematics and Philosophy in late Antiquity* (Oxford 1989).

32 See Christian Guérard, 'Parménide d'Elée selon les Néoplatoniciens', forthcoming.

33 Simplicius *in Phys.* 28,32-29,5; 640,12-18. Such thinkers as Epicurus and the Sceptics, however, were not subject to harmonisation.

34 See the literature in n. 2 above.

35 ap. Elian *in Cat.* 107,24-6.

the commentator to find a unitary theme or scope (*skopos*) for the treatise. The arrangements for late commentaries on Plato are similar. Since the Plato commentaries form part of a single curriculum they should be studied alongside those on Aristotle. Here the situation is easier, not only because the extant corpus is very much smaller, but also because it has been comparatively well served by French and English translators.[36]

Given the theological motive of the curriculum and the pressure to harmonise Plato with Aristotle, it can be seen how these commentaries are a major source for Neoplatonist ideas. This in turn means that it is not safe to extract from them the fragments of the Presocratics, or of other authors, without making allowance for the Neoplatonist background against which the fragments were originally selected for discussion. For different reasons, analogous warnings apply to fragments preserved by the pre-Neoplatonist commentator Alexander.[37] It will be another advantage of the present translations that they will make it easier to check the distorting effect of a commentator's background.

Although the Neoplatonist commentators conflate the views of Aristotle with those of Neoplatonism, Philoponus alludes to a certain convention when he quotes Plutarchus expressing disapproval of Alexander for expounding his own philosophical doctrines in a commentary on Aristotle.[38] But this does not stop Philoponus from later inserting into his own commentaries on the *Physics* and *Meteorology* his arguments in favour of the Christian view of Creation. Of course, the commentators also wrote independent works of their own, in which their views are expressed independently of the exegesis of Aristotle. Some of these independent works will be included in the present series of translations.

The distorting Neoplatonist context does not prevent the commentaries from being incomparable guides to Aristotle. The introductions to Aristotle's philosophy insist that commentators must have a minutely detailed knowledge of the entire Aristotelian corpus, and this they certainly have. Commentators are also enjoined neither to accept nor reject what Aristotle says too readily, but to consider it in depth and without partiality. The commentaries draw one's attention to hundreds of phrases, sentences and ideas in Aristotle, which one could easily have passed over, however often one read him. The scholar who makes the right allowance for the distorting context will learn far more about Aristotle than he would be likely to on his own.

The relations of Neoplatonist commentators to the Christians were subtle. Porphyry wrote a treatise explicitly against the Christians in 15 books, but an order to burn it was issued in 448, and later Neoplatonists

[36] English: Calcidius *in Tim.* (parts by van Winden; den Boeft); Iamblichus fragments (Dillon); Proclus *in Tim.* (Thomas Taylor); Proclus *in Parm.* (Dillon); Proclus *in Parm.*, end of 7th book, from the Latin (Klibansky, Labowsky, Anscombe); Proclus *in Alcib. 1* (O'Neill); Olympiodorus and Damascius *in Phaedonem* (Westerink); Damascius *in Philebum* (Westerink); *Anonymous Prolegomena to Platonic Philosophy* (Westerink). See also extracts in Thomas Taylor, *The Works of Plato*, 5 vols. (1804). French: Proclus *in Tim.* and *in Rempublicam* (Festugière); *in Parm.* (Chaignet); Anon. *in Parm.* (P. Hadot); Damascius *in Parm.* (Chaignet).

[37] For Alexander's treatment of the Stoics, see Robert B. Todd, *Alexander of Aphrodisias on Stoic Physics* (Leiden 1976), 24-9.

[38] Philoponus *in DA* 21,20-3.

were more circumspect. Among the last commentators in the main group, we have noted several Christians. Of these the most important were Boethius and Philoponus. It was Boethius' programme to transmit Greek learning to Latin-speakers. By the time of his premature death by execution, he had provided Latin translations of Aristotle's logical works, together with commentaries in Latin but in the Neoplatonist style on Porphyry's *Isagôgê* and on Aristotle's *Categories* and *de Interpretatione*, and interpretations of the *Prior* and *Posterior Analytics, Topics* and *Sophistici Elenchi*. The interruption of his work meant that knowledge of Aristotle among Latin-speakers was confined for many centuries to the logical works. Philoponus is important both for his proofs of the Creation and for his progressive replacement of Aristotelian science with rival theories, which were taken up at first by the Arabs and came fully into their own in the West only in the sixteenth century.

Recent work has rejected the idea that in Alexandria the Neoplatonists compromised with Christian monotheism by collapsing the distinction between their two highest deities, the One and the Intellect. Simplicius (who left Alexandria for Athens) and the Alexandrians Ammonius and Asclepius appear to have acknowledged their beliefs quite openly, as later did the Alexandrian Olympiodorus, despite the presence of Christian students in their classes.[39]

The teaching of Simplicius in Athens and that of the whole pagan Neoplatonist school there was stopped by the Christian Emperor Justinian in 529. This was the very year in which the Christian Philoponus in Alexandria issued his proofs of Creation against the earlier Athenian Neoplatonist Proclus. Archaeological evidence has been offered that, after their temporary stay in Ctesiphon (in present-day Iraq), the Athenian Neoplatonists did not return to their house in Athens, and further evidence has been offered that Simplicius went to Ḥarrān (Carrhae), in present-day Turkey near the Iraq border.[40] Wherever he went, his commentaries are a treasure house of information about the preceding thousand years of Greek philosophy, information which he painstakingly recorded after the closure in Athens, and which would otherwise have been lost. He had every reason to feel bitter about Christianity, and in fact he sees it and Philoponus, its representative, as irreverent. They deny the divinity of the heavens and prefer the physical relics of dead martyrs.[41] His own commentaries by

[39] For Simplicius, see I. Hadot, *Le Problème du Néoplatonisme Alexandrin: Hiéroclès et Simplicius* (Paris 1978); for Ammonius and Asclepius, Koenraad Verrycken, *God en Wereld in de Wijsbegeerte van Ioannes Philoponus*, Ph.D. Diss. (Louvain 1985); for Olympiodorus, L.G. Westerink, *Anonymous Prolegomena to Platonic Philosophy* (Amsterdam 1962).

[40] Alison Frantz, 'Pagan philosophers in Christian Athens', *Proceedings of the American Philosophical Society* 119 (1975), 29-38; M. Tardieu, 'Témoins orientaux du *Premier Alcibiade* à Ḥarrān et à Nag 'Hammādi', *Journal Asiatique* 274 (1986); id., 'Les calendriers en usage à Ḥarrān d'après les sources arabes et le commentaire de Simplicius à la *Physique* d'Aristote', in I. Hadot (ed.), *Simplicius, sa vie, son oeuvre, sa survie* (Berlin 1987), 40-57; id., *Coutumes nautiques mésopotamiennes chez Simplicius*, in preparation. The opposing view that Simplicius returned to Athens is most fully argued by Alan Cameron, 'The last days of the Academy at Athens', *Proceedings of the Cambridge Philological Society* 195, n.s. 15 (1969), 7-29.

[41] Simplicius *in Cael.* 26,4-7; 70,16-18; 90,1-18; 370,29-371,4. See on his whole attitude Philippe Hoffmann, 'Simplicius' polemics', in Richard Sorabji (ed.), *Philoponus and the Rejection of Aristotelian Science* (London and Ithaca, N.Y. 1987).

contrast culminate in devout prayers.

Two collections of articles by various hands have been published, to make the work of the commentators better known. The first is devoted to Philoponus;[42] the second is about the commentators in general, and goes into greater detail on some of the issues briefly mentioned here.[43]

[42] Richard Sorabji (ed.), *Philoponus and the Rejection of Aristotelian Science* (London and Ithaca, N.Y. 1987).

[43] Richard Sorabji (ed.), *Aristotle Transformed: the ancient commentators and their influence* (London and Ithaca, N.Y. 1990). The lists of texts and previous translations of the commentaries included in Wildberg, *Philoponus Against Aristotle on the Eternity of the World* (pp.12ff.) are not included here. The list of translations should be augmented by: F.L.S. Bridgman, Heliodorus (?) in *Ethica Nicomachea*, London 1807.

I am grateful for comments to Henry Blumenthal, Victor Caston, I. Hadot, Paul Mercken, Alain Segonds, Robert Sharples, Robert Todd, L.G. Westerink and Christian Wildberg.

# English-Greek Glossary

absurd, *atopos*
abyss, *buthos*
account, *logos* (non-technical)
actuality, *energeia*
actually, *energeiâi*
aether, *aithêr*
aethereal, *aithêrios*
be affected, *paskhein*
air, *aer*
alteration, *alloiôsis*
animate, *empsukhos*
attribute, *sumbebêkos*

to be, *einai*
become, *ginesthai*
beginning, *arkhê*
body, *sôma*
breadth, *platos*
bulk, *onkos*

cardinal points, *kentra*
cause, *aitia*
change, *metabolê*
changeable, *pathêtos*
characteristic (adj.), *oikheios*
coherence, *sumptôsis*
be coincident with, *sunarmozein*
composed of elements, *stoikheiôtos*
composite, *merismos*
concentration, *sunairesis*
container, *periekhon*
continuous, *sunekhês*
coplacement, *suntopôsis*
copy, *eikôn*
corporeal, *sômatikos*

dance, *khoreia*
dance around, *khoreuein peri*
decline, *huphesis*
degeneration, *ekstasis*
depth, *bathos*
descend, *hupobainein*
descent, *hupobasis*
differentiation, *diakrisis*
dimension, *diastêma*, *diastasis*
distinguishing mark, *diaphora*
diverge, *parallatein*
divergence, *parallaxis*
divine, *theios*
divisible, *diairetos*, *meristos*
duration, *diastêma*

earth, *gê*
element, *stoikheion*
emanation, *probolê*
enmattered, *enulos*
essence, *ousia*, *to einai*
essential, *kat' ousian*, *ousiôdês*
essentially joined with, *sunousiômenos*
eternal, *aiônios*
eternity, *aiôn*
ever, *aei*
everlasting, *aei*
exchange of place, *metabasis*
existence, *to einai*
existent, *to on*
explanation, *aitia*
extended, *diastatos*
extension, *diastêma*

firmaments, *stereômata*
fixed, *aplanês*
flow (of time), *rhoê*
flow out, *ekkhein*
form, *eidos*
future, *mellon*

generated, *genêtos*
generation, *genesis*
generative, *genesiourgios*
guess, *topazein*

hard to distinguish, *dusdiakritos*
heavens, *ouranos*
here, *entautha*, *enthade*

iconic, *eikonikos*
immaterial, *aülos*
immobile, *akinêtos*
impart, *didonai*
impassive, *apathês*
imperishable, *aphthartos*
incomposite, *ameristos*
incorporeal, *asômatos*
indeterminate, *aoristos*
indivisible, *adiairetos*
infinite in power, *apeirodunamos*
intellect, *nous*
intellective, *noeros*
intelligible, *noêtos*
intermediate, *mesotês*
interval, *diastêma*

leap, *halma*
length, *mêkos*
life, *zôê*
lifeless, *apsukhos*
limit, *peras*
living thing, *zôios*
locomotion, *phora*

manifold, *plêthos*
material, *hulaios*
matter, *hulê*
measure, *metron*
measurer, *metroun*
motion, *kinêsis*
mould, *tupos*

naturally joined, *sumpephukos*, *sumphutos*
nature, *phusis*
now, *nun*
number, *arithmos*

one, *hen*

part, *meros*
participation, *methexis*
a particular, *tode ti*
past, *parelthon*
pattern, *paradeigma*
perceptible, *aisthêtos*
perpetual, *aïdios*
penetration, *diïxis*
perishable, *phthartos*
place, *topos*
position, *thesis*
possible, *dunatos*
postulate, *axiôma*
potentially, *dunamei*
pre-eminently, *proêgoumenôs*
preliminary outline, *proüpographê*
present, *enestôs*, *enestêkos*
principle, *logos*
problem, *aporia*
proceed, *proerkhesthai*, *proïenai*
primal source, from the, *pêgaios*
process, *kinêsis*
procession, *proödos*
proper, *oikeios*
puzzle, *aporia*

quality, *poion*, *poiotês*
quantity, *poson*, *posotês*

receptacle, *hupodokhê*
rest, *êremia*
revert, *epistrephein*
rise above, *huperanekhein*

sideturn, *ektropê*
size, *megethos*
soul, *psukhê*
space, *khôra*
species, *eidos*
straight, *euthus*
striving, *anatasis*
sublunary, *hupo selênên*
substance, *ousia*
substrate, *hupokeimenon*
supratemporal, *huper khronon*
synchronic, *athroös*

tangible, *haptos*
there, *ekei*, *ekeithen*
theory, *hupothesis*
time, *khronos*
transcendent, *exeirêmenos*
true reality, *to ontôs on*
truly, *ontôs*

undifferentiated, *adiaphoros*
unfolding, *anelixis*
ungenerated, *agenêtos*
unifying, *sunagôgos*
unity, *henôsis*, *hen*
universal, *katholou*
unlimited, *apeiros*

vessel, *angeion*

water, *hudôr*

# Greek-English Index

References are to the page and line numbers of the *CAG* edition, which appear in the margin of this translation.

# Subject Index

References are to the page and line numbers of the *CAG* edition, which appear in the margin of this translation.